Time to face all the facts with CGP!

Make sure you know all the key concepts for
Higher GCSE Maths with this CGP Knowledge Organiser!

We've boiled every topic down to the essentials, with step-by-step
methods and worked examples to help it all sink in.

There's also a matching Knowledge Retriever book that'll test
you on every page — perfect for making sure you know it all!

CGP — still the best! ☺

Our sole aim here at CGP is to produce the highest quality books —
carefully written, immaculately presented and dangerously close to being funny.

Then we work our socks off to get them out to you
— at the cheapest possible prices.

Contents

Section 6 — Pythagoras and Trigonometry

Section 7 — Probability and Statistics

Published by CGP.
From original material by Richard Parsons.

Editors: Sarah George, Sharon Keeley-Holden, Samuel Mann, Sean McParland, Caley Simpson.

With thanks to Judith Hayes and Glenn Rogers for the proofreading.
With thanks to Emily Smith for the copyright research.

Printed by Elanders Ltd, Newcastle upon Tyne.
Clipart from Corel®

Types of Number and BODMAS

Seven Types of Numbers

		Definition	Examples
1	INTEGER	Whole number	–16, 0, 2, 547
2	RATIONAL	Can be written as a fraction	$0 \left(= \frac{0}{1}\right)$, $0.44... \left(= \frac{4}{9}\right)$
3	IRRATIONAL	Can't be written as a fraction — never-ending, non-repeating decimals	$\sqrt{2}$, $5\sqrt{3}$, π
4	NEGATIVE	Less than zero	–21, –3.6, –0.01
5	MULTIPLE	In a number's times table (or beyond)	Of 3: 3, 6, 15, 42
6	FACTOR	Divides into a number	Of 10: 1, 2, 5, 10
7	PRIME	Only factors are itself and 1	2, 3, 17, 43

1 is NOT prime.

Two Rules for Dealing with Negative Numbers

1 Signs the same: positive

$+12 - -5 = 12 + 5 = +17$

$(-5)^2 = -5 \times -5 = +25$

2 Different signs: negative

$+15 + -13 = 15 - 13 = 2$

$-121 \div +11 = -11$

Use these rules when multiplying or dividing, or when two signs are together.

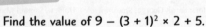

BODMAS

BODMAS gives the order of operations:

1 Brackets

'Other' is things like squaring.

2 Other

3 Division and Multiplication

4 Addition and Subtraction

EXAMPLE

Find the value of $9 - (3 + 1)^2 \times 2 + 5$.

$9 - (3 + 1)^2 \times 2 + 5$

① $= 9 - 4^2 \times 2 + 5$

② $= 9 - 16 \times 2 + 5$

③ $= 9 - 32 + 5$ — Work left to right when there's only addition

④ $= -23 + 5$ — and subtraction.

$= -18$

Multiples and Factors

Four Steps to Find Factors

1. List factors in pairs, starting with 1 × the number, then 2 ×, etc.

2. Cross out pairs that don't divide exactly.

3. Stop when a number is repeated.

4. Write factors out clearly.

Find all the factors of 20.

1. 1 × 20
   2 × 10
2. ~~3 ×~~
   4 × 5
3. 5 × 4
4. So the factors of 20 are:
   1, 2, 4, 5, 10, 20

Finding Prime Factors

PRIME FACTORISATION — writing a number as its prime factors multiplied together.

Three steps to use a Factor Tree:

1. Put the number at the top and split into factors.

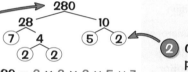

2. Circle each prime.

3. When only primes are left, write them in order.

$280 = 2 \times 2 \times 2 \times 5 \times 7$
$= 2^3 \times 5 \times 7$

Lowest Common Multiple (LCM)

LCM — the smallest number that divides by all numbers in question.

Find it from prime factors in two steps:

1. List all prime factors in either number.

2. Multiply together.

EXAMPLE

Find the LCM of 8 and 14.

$8 = 2 \times 2 \times 2$
$14 = 2 \times 7$

If a factor appears more than once in any number, list it that many times.

1. 2, 2, 2, 7
2. $2 \times 2 \times 2 \times 7 = 56$

Highest Common Factor (HCF)

HCF — the biggest number that divides into all numbers in question.

Find it from prime factors in two steps:

1. List all prime factors that are in both numbers.

2. Multiply together.

EXAMPLE

Find the HCF of 36 and 90.

$36 = 2 \times 2 \times 3 \times 3$
$90 = 2 \times 3 \times 3 \times 5$

1. 2, 3, 3   2. $2 \times 3 \times 3 = 18$

You can also find LCM/HCF by listing the multiples/factors of both numbers and taking the smallest/biggest number that appears in both lists.

Fractions

Simplifying Fractions

To simplify, divide top and bottom by the same number until they won't divide any more.

$$\frac{30}{45} \xrightarrow[\div 5]{\div 5} \frac{6}{9} \xrightarrow[\div 3]{\div 3} \frac{2}{3}$$

$$\frac{42}{70} \xrightarrow[\div 7]{\div 7} \frac{6}{10} \xrightarrow[\div 2]{\div 2} \frac{3}{5}$$

Mixed Numbers and Improper Fractions

MIXED NUMBER — has integer part and fraction part, e.g. $2\frac{1}{3}$.

IMPROPER FRACTION — has numerator larger than denominator, e.g. $\frac{7}{5}$.

To write mixed numbers as improper fractions:

1 Write as an addition.

2 Turn integer part into a fraction.

3 Add together.

$$2\frac{3}{4} = 2 + \frac{3}{4} = \frac{8}{4} + \frac{3}{4} = \frac{11}{4}$$
1   **2**   **3**

To write improper fractions as mixed numbers:

1 Divide top by bottom.

2 Answer is whole number part, remainder goes on top of fraction part.

1 $17 \div 3 = 5$ remainder 2

2 So $\frac{17}{3} = 5\frac{2}{3}$

Multiplying and Dividing

1 Rewrite any mixed numbers as fractions.

If multiplying

If dividing

2 Turn 2nd fraction upside down. Change ÷ to ×.

3 Cancel down with common factors.

4 Multiply tops and bottoms separately.

EXAMPLE

Find $1\frac{3}{5} \div \frac{6}{7}$.

1 $1\frac{3}{5} \div \frac{6}{7} = \frac{8}{5} \div \frac{6}{7}$

2 $= \frac{{}^4 8}{5} \times \frac{7}{6_3}$

3 $= \frac{4}{5} \times \frac{7}{3}$

4 $= \frac{4 \times 7}{5 \times 3} = \frac{28}{15}$

Common Denominators

Use to order, add or subtract fractions.

Find a number that all denominators divide into — the LCM is best.

EXAMPLE

Put $\frac{11}{6}, \frac{17}{12}, \frac{7}{4}$ in descending order.

LCM of 6, 12, 4 is 12.

$$\frac{11}{6} \xrightarrow[\times 2]{\times 2} \frac{22}{12} \qquad \frac{7}{4} \xrightarrow[\times 3]{\times 3} \frac{21}{12}$$

$$\frac{22}{12} > \frac{21}{12} > \frac{17}{12}$$

So $\frac{11}{6}, \frac{7}{4}, \frac{17}{12}$

Fractions, Decimals and Percentages

Adding and Subtracting Fractions

1 Make denominators the same.

2 Add/subtract the numerators only.

EXAMPLE

Find $1\frac{1}{3} - \frac{5}{8}$.

$1\frac{1}{3} - \frac{5}{8} = \frac{4}{3} - \frac{5}{8} = \frac{32}{24} - \frac{15}{24}$ ①

Rewrite any mixed numbers.

② $= \frac{32-15}{24} = \frac{17}{24}$

Fractions of Amounts

To find a fraction of a number:

1 Divide the number by the denominator.

2 Multiply by the numerator.

$\frac{7}{12}$ of 240 = (240 ÷ 12) × 7 ①

② = 20 × 7 = 140

Multiply then divide if it's easier.

To write one number as a fraction of another:

1 Write the 1st number over the 2nd.

2 Cancel down.

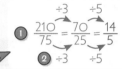

① $\frac{210}{75} = \frac{70}{25} = \frac{14}{5}$

② ÷3   ÷5

Common Conversions

Fractions, decimals and percentages are all proportions.
You can convert between them.

Fraction	Decimal	Percentage	Fraction	Decimal	Percentage
$\frac{1}{2}$	0.5	50%	$\frac{1}{10}$	0.1	10%
$\frac{1}{4}$	0.25	25%	$\frac{1}{5}$	0.2	20%
$\frac{3}{4}$	0.75	75%	$\frac{1}{8}$	0.125	12.5%
$\frac{1}{3}$	0.3333...	$33\frac{1}{3}\%$	$\frac{3}{8}$	0.375	37.5%
$\frac{2}{3}$	0.6666...	$66\frac{2}{3}\%$	$\frac{5}{2}$	2.5	250%

How to Convert

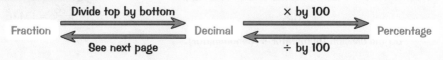

Divide top by bottom →

× by 100 →

Fraction   Decimal   Percentage

See next page

÷ by 100

Terminating and Recurring Decimals

Converting Terminating Decimals

TERMINATING DECIMALS — are finite (come to an end), e.g. 0.7 and 2.618. When simplified, denominators have only 2 and 5 as prime factors.

Three steps to write as fractions:

1. Put the digits after the decimal point as the numerator.

2. Count the decimal places and put a power of 10 with that many zeros as the denominator.

3. Cancel down.

EXAMPLE

Write these decimals as fractions in their simplest forms:

a) 0.308

① $0.308 = \dfrac{308}{} = $ ② $\dfrac{308}{1000}$

③ $= \dfrac{154}{500} = \dfrac{77}{250}$ 3 decimal places, so use $10^3 = 1000$.

b) 0.0015

① $0.0015 = \dfrac{15}{}$ ② $= \dfrac{15}{10\,000}$

③ $= \dfrac{3}{2000}$ 4 decimal places, so use $10^4 = 10\,000$.

.6 .6 .6

Converting Recurring Decimals

RECURRING DECIMALS — have a pattern of numbers that repeats forever.
Repeated bit is marked with dots. One dot: that digit is repeated, e.g. $0.1\dot{6} = 0.1666...$
Two dots: everything from the first to second is repeated, e.g. $0.\dot{1}8\dot{7} = 0.187187...$

To write a recurring decimal as a fraction:

1. Name the decimal r.

2. Multiply r by a power of 10 to get any non-repeating parts past the decimal point.

3. Multiply by a power of 10 again to get one full repeated lump past the decimal point.

4. Subtract to get rid of the decimal part.

5. Divide and cancel to find r.

EXAMPLE

Write $0.5\dot{7}$ as a fraction.

① Let $r = 0.5\dot{7}$

② $10r = 5.\dot{7}$

③ $100r = 57.\dot{7}$

④ $\begin{array}{r} 100r = 57.\dot{7} \\ -\quad 10r = 5.\dot{7} \\ \hline 90r = 52 \end{array}$

⑤ $r = \dfrac{52}{90} = \dfrac{26}{45}$

To write a fraction as a recurring decimal:

Find an equivalent fraction with all nines in the denominator — the numerator is the recurring part.

OR

Do the division (numerator ÷ denominator).

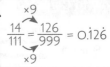

$\dfrac{14}{111} \overset{\times 9}{\underset{\times 9}{=}} \dfrac{126}{999} = 0.\dot{1}2\dot{6}$

Rounding and Estimating

Rounding to Decimal Places (d.p.) and Significant Figures (s.f.)

Digit after the last digit is the decider:

- If decider is 5 or more, round last digit up.
- If decider is 4 or less, leave last digit as is.

To find significant figures:

1 The first non-zero digit is the first s.f.

2 Each following digit (zero or non-zero) is another s.f.

After rounding, fill in with zeros up to, not beyond, the decimal point.

18.074 rounded to:

3 s.f.: Last digit = 0, so decider = 7 — round up to 18.1

2 d.p.: Last digit = 7, so decider = 4 — leave as 18.07

1 s.f.: Last digit = 1, so decider = 8 — round up to 20 ⟵ Fill in with a 0.

Estimating Calculations and Square Roots

To estimate calculations, round all numbers to either 1 or 2 s.f.

$$\frac{20.2 \times 2.87}{5.913} \approx \frac{20 \times 3}{6} = \frac{60}{6} = 10$$

Two steps to estimate square roots:

1 Find a square number on each side of the given number.

2 Decide which it's closer to, then estimate the digit after the decimal point.

EXAMPLE

Estimate the value of $\sqrt{68}$ to 1 d.p.

1 $64 (= 8^2) < 68 < 81 (= 9^2)$

2 68 is closer to 64 than 81, so $\sqrt{68}$ is closer to 8 than 9. $\sqrt{68} \approx 8.2$

Upper and Lower Bounds

LOWER BOUND $\leq$ actual value $<$ UPPER BOUND

Rounded value – half a unit	$\leq$	actual value	$<$	Rounded value + half a unit

Truncated value	$\leq$	actual value	$<$	Truncated value + 1 whole unit

To find max and min values for a calculation:

1 Find bounds for each number.

2 Pick bounds to use for each operation.

EXAMPLE

To 1 d.p., $x = 1.4$ and $y = 3.7$. What are the maximum and minimum values of $x \times y$?

1 $1.35 \leq x < 1.45$ ⟵ 1 d.p. is 0.1, so half of this is 0.05.
$3.65 \leq y < 3.75$

2 $\max(x \times y) = \max(x) \times \max(y)$
$= 1.45 \times 3.75 = 5.4375$

$\min(x \times y) = \min(x) \times \min(y)$
$= 1.35 \times 3.65 = 4.9275$

$\max(a \div b) = \max(a) \div \min(b)$
$\min(a \div b) = \min(a) \div \max(b)$

Section 1 — Number

Standard Form

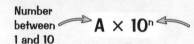

Number between 1 and 10 $\Rightarrow$ $A \times 10^n$ $\Leftarrow$ Number of places the decimal point moves — positive for big numbers, negative for small numbers

EXAMPLE What is 70.6 million in standard form?

70.6 million = 70.6 × 1 000 000

= 70 600 000.0 — Count how far the decimal point moves.

= 7.06×10^7 — Big number, so positive

EXAMPLE Express 5.129×10^{-4} as an ordinary number.

Negative, so small number

00005.129 × 10^{-4} — Move the decimal point by this many places.

= 0.0005129

Three Steps to Multiply or Divide

1. Rearrange so the front numbers and powers of 10 are together.

2. Multiply/divide the front numbers. Use power rules to multiply/divide the powers of 10.

3. Put the answer in standard form.

EXAMPLE

Find $(8.15 \times 10^7) \times (4 \times 10^{-3})$. Give your answer in standard form.

$(8.15 \times 10^7) \times (4 \times 10^{-3})$

1. $= (8.15 \times 4) \times (10^7 \times 10^{-3})$

2. $= 32.6 \times 10^{7 + -3}$ — Add the powers.

$= 32.6 \times 10^4$

3. $= 3.26 \times 10 \times 10^4$

$= 3.26 \times 10^5$

Three Steps to Add or Subtract

1. Rewrite so the powers of 10 are the same.

2. Add/subtract front numbers.

3. Put the answer in standard form.

EXAMPLE

Find $(3.4 \times 10^{-6}) + (9.7 \times 10^{-5})$. Give your answer in standard form.

$(3.4 \times 10^{-6}) + (9.7 \times 10^{-5})$

1. $= (0.34 \times 10 \times 10^{-6}) + (9.7 \times 10^{-5})$

$= (0.34 \times 10^{-5}) + (9.7 \times 10^{-5})$

2. $= (0.34 + 9.7) \times 10^{-5}$

$= 10.04 \times 10^{-5}$ — Not in standard form yet — front number is bigger than 10.

3. $= 1.004 \times 10 \times 10^{-5}$

$= 1.004 \times 10^{-4}$

Algebra Basics

Algebraic Notation

Only q
is cubed
— not p.

Notation	Meaning
abc	a × b × c
$\frac{a}{b}$	a ÷ b
pq³	p × q × q × q
(mn)²	m × m × n × n
x(y − z)³	x × (y − z) × (y − z) × (y − z)

Brackets mean
both m and n
are squared.

Things like −4²
are unclear. Write
either (−4)² = 16 or
−(4²) = −16 instead.

Collecting Like Terms

TERM — a collection of numbers, letters
and brackets, all multiplied/divided together.

1 Put bubbles around each term.

2 Move bubbles so
like terms are together.

3 Combine like terms.

EXAMPLE

Simplify $7a + 2 − 3a + 5$.

① $7a$ $+2$ $−3a$ $+5$ — Put the
+/− sign in
each bubble.

② $= 7a$ $−3a$ $+2$ $+5$

③ $= 4a + 7$

Ten Rules for Powers

These are only true for
powers of the same number.

1 Multiplying — ADD powers:
e.g. $a^2 \times a^5 = a^7$

2 Dividing — SUBTRACT powers:
e.g. $b^5 \div b^3 = b^2$

3 Raising one power to another
— MULTIPLY powers: e.g. $(p^2)^4 = p^8$

4 Anything to the power of 1 is ITSELF:
e.g. $x^1 = x$

5 Anything to the power of 0 is 1:
e.g. $y^0 = 1$

6 1 to the power of anything is still 1:
e.g. $1^x = 1$

7 Apply powers to the TOP
and BOTTOM of fractions: e.g. $\left(\frac{m}{n}\right)^2 = \frac{m^2}{n^2}$

8 NEGATIVE powers — flip it over,
then make the power positive.

$$5^{-2} = \frac{1}{5^2} = \frac{1}{25}$$

9 FRACTIONAL powers are roots —
e.g. power of $\frac{1}{2}$ is a square root,
power of $\frac{1}{3}$ is a cube root, etc.

$$16^{\frac{1}{4}} = \sqrt[4]{16} = 2$$

10 TWO-STAGE FRACTIONAL powers
— do the root, then the power.

$$27^{\frac{2}{3}} = \left(27^{\frac{1}{3}}\right)^2 = 3^2 = 9$$

Expanding Brackets and Factorising

Expanding Brackets

Multiply everything inside the bracket by everything outside the bracket.

$$2x(5 - 3y) = (2x \times 5) + (2x \times -3y)$$
$$= 10x - 6xy$$

The FOIL method for double brackets:

- Multiply First terms of each bracket.
- Multiply Outside terms together.
- Multiply Inside terms together.
- Multiply Last terms of each bracket.

$$(m - 6)(3m + 4)$$

$$= (m \times 3m) + (m \times 4)$$
$$+ (-6 \times 3m) + (-6 \times 4)$$
$$= 3m^2 + 4m - 18m - 24$$
$$= 3m^2 - 14m - 24$$

To multiply out triple brackets, multiply two together as normal, then multiply the result by the third bracket.

Factorising Expressions

FACTORISING — putting brackets back in.

1. Take out biggest number that goes into all terms.

2. Take out highest power of each letter that goes into all terms.

3. Open bracket and fill in what's needed to reproduce the original terms.

4. Check your answer.

$$3b^2 - 6ab = 3b(b - 2a)$$

④ $3b(b - 2a) = 3b \times b - 3b \times 2a$
$$= 3b^2 - 6ab$$

The bits put in front of the bracket are the common factors.

The Difference of Two Squares (D.O.T.S.)

D.O.T.S. — 'one thing squared' take away 'another thing squared'.

Use this rule for factorising: $a^2 - b^2 = (a + b)(a - b)$

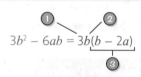

The difference? The colour and about 45°...

EXAMPLE

Factorise $5p^2 - 20q^2$.

$$5p^2 - 20q^2 = 5(p^2 - 4q^2) = 5(p + 2q)(p - 2q)$$

You might need to take out a factor to get it in the form $a^2 - b^2$.

Surds and Solving Equations

Six Rules for Manipulating Surds

① $\sqrt{a} \times \sqrt{b} = \sqrt{a \times b}$

$\sqrt{5} \times \sqrt{3} = \sqrt{15}$

⑤ $(a + \sqrt{b})(a - \sqrt{b}) = a^2 - b$

$(4 + \sqrt{7})(4 - \sqrt{7}) = 16 + 4\sqrt{7} - 4\sqrt{7} - 7$
$= 16 - 7 = 9$

② $\dfrac{\sqrt{a}}{\sqrt{b}} = \sqrt{\dfrac{a}{b}}$

$\dfrac{\sqrt{27}}{\sqrt{3}} = \sqrt{\dfrac{27}{3}} = \sqrt{9} = 3$

⑥ $\dfrac{a}{\sqrt{b}} = \dfrac{a\sqrt{b}}{b}$

$\dfrac{3}{\sqrt{5}} = \dfrac{3}{\sqrt{5}} \times \dfrac{\sqrt{5}}{\sqrt{5}} = \dfrac{3\sqrt{5}}{5}$

This is known as 'rationalising the denominator'.

③ $\sqrt{a} + \sqrt{b}$ — do nothing.
(Definitely NOT $\sqrt{a+b}$)

④ $(a + \sqrt{b})^2 = a^2 + 2a\sqrt{b} + b$

$(6 + \sqrt{2})^2 = (6 + \sqrt{2})(6 + \sqrt{2})$
$= 36 + 12\sqrt{2} + 2$
$= 38 + 12\sqrt{2}$

EXAMPLE

Write $\sqrt{54} + \sqrt{150} - \sqrt{24}$ in the form $a\sqrt{6}$.

$\sqrt{54} = \sqrt{9 \times 6} = \sqrt{9} \times \sqrt{6} = 3\sqrt{6}$
$\sqrt{150} = \sqrt{25 \times 6} = \sqrt{25} \times \sqrt{6} = 5\sqrt{6}$
$\sqrt{24} = \sqrt{4 \times 6} = \sqrt{4} \times \sqrt{6} = 2\sqrt{6}$
$3\sqrt{6} + 5\sqrt{6} - 2\sqrt{6} = 6\sqrt{6}$

Six Steps to Solve Equations

① Get rid of fractions.

② Multiply out brackets.

③ Put x terms on one side, numbers on the other.

④ Reduce to the form Ax = B.

⑤ Divide by A to get 'x = ...'.

⑥ If you have 'x² = ...' instead, square root both sides.

You can ignore any steps that don't apply to the equation.

EXAMPLE

Solve $\dfrac{3}{x-2} = \dfrac{2}{3x+1}$.

① $3(3x + 1) = 2(x - 2)$
② $9x + 3 = 2x - 4$
③ $9x - 2x = -4 - 3$
④ $7x = -7$
⑤ $x = -1$ — There's no x^2, so stop at Step 5.

EXAMPLE

Solve $x(5x) - 13 = 7$.

② $5x^2 - 13 = 7$
③ $5x^2 = 7 + 13$
④ $5x^2 = 20$
⑤ $x^2 = 4$
⑥ $x = \pm 2$

Taking the square root gives a positive and a negative solution.

Rearranging Formulas

Seven Steps for Rearranging Formulas

① Get rid of square roots.

② Get rid of fractions.

③ Multiply out brackets.

④ Put subject terms on one side, non-subject terms on the other.

⑤ Reduce to the form $Ax = B$ (where x is the subject).

⑥ Divide by A to get 'x = ...'.

⑦ If you're left with '$x^2 = ...$', square root both sides.

A and B could be numbers, letters or a mix of both.

If the Subject is in a Fraction

EXAMPLE

Make p the subject of $q = \dfrac{7p - 3}{5}$.

② $5q = 7p - 3$

④ $7p = 5q + 3$

⑤ This is in the form $Ap = B$.

⑥ $p = \dfrac{5q + 3}{7}$

No square roots or brackets, so ignore Steps 1 and 3.

If the Subject Appears Twice

You'll need to factorise, usually at Step 5.

EXAMPLE

Make m the subject of $n = \dfrac{m}{m - 3}$.

② $n(m - 3) = m$

③ $mn - 3n = m$

This is where you factorise — m is a common factor.

④ $mn - m = 3n$

⑤ $m(n - 1) = 3n$

⑥ $m = \dfrac{3n}{n - 1}$

No square roots, so ignore Step 1.

If there's a Square or Square Root

EXAMPLE

Make r the subject of $s^2 = 9 - 3r^2$.

④ $3r^2 = 9 - s^2$

⑤ This is in the form $Ar^2 = B$.

⑥ $r^2 = \dfrac{9 - s^2}{3}$

⑦ $r = \pm\sqrt{\dfrac{9 - s^2}{3}}$

No square roots, fractions or brackets, so ignore Steps 1-3.

EXAMPLE

Make a the subject of $2b + 1 = \sqrt{4a - 3}$.

① $(2b + 1)^2 = 4a - 3$

③ $4b^2 + 4b + 1 = 4a - 3$

④ $4a = 4b^2 + 4b + 4$

⑤ This is in the form $Aa = B$.

⑥ $a = b^2 + b + 1$

No fractions, so ignore Step 2.

Factorising Quadratics

Quadratic Equations

Standard form of a quadratic equation: $ax^2 + bx + c = 0$

a, b and c can be any number.

To FACTORISE — put it into two brackets.
To SOLVE — find the values of x that make each bracket equal to 0.

Factorising when a = 1

1 Rearrange to $x^2 + bx + c = 0$.

2 Write two brackets: $(x \quad)(x \quad) = 0$

3 Find two numbers that multiply to give 'c' AND add/subtract to give 'b'.

4 Fill in + or – signs.

5 Check by expanding brackets.

6 Solve the equation.

EXAMPLE

Solve $x^2 - 6x = -8$.

1 $x^2 - 6x + 8 = 0$

2 $(x \quad)(x \quad) = 0$

3 Factor pairs of 8: 1×8 or 2×4
$2 + 4 = 6$, so you need 2 and 4.

4 $(x - 2)(x - 4) = 0$

5 $(x - 2)(x - 4) = x^2 - 4x - 2x + 8$
$\qquad\qquad\qquad = x^2 - 6x + 8$

6 $(x - 2) = 0 \Rightarrow x = 2$
$(x - 4) = 0 \Rightarrow x = 4$

Factorising when a is not 1

1 Rearrange to $ax^2 + bx + c = 0$.

2 Write two brackets where the first terms multiply to give 'a'.

3 Find pairs of numbers that multiply to give 'c'.

4 Test each pair in both brackets to find one that adds/subtracts to give 'bx'.

5 Fill in + or – signs.

6 Check by expanding brackets.

7 Solve the equation.

EXAMPLE

Solve $2x^2 + x - 6 = 0$. **1** This is in the standard format.

2 $(2x \quad)(x \quad) = 0$

3 Factor pairs of 6: 1×6 or 2×3

4 $(2x \quad 1)(x \quad 6) \rightarrow 12x$ and x
$(2x \quad 6)(x \quad 1) \rightarrow 2x$ and $6x$
$(2x \quad 2)(x \quad 3) \rightarrow 6x$ and $2x$
$(2x \quad 3)(x \quad 2) \rightarrow 4x$ and $3x$

5 $(2x - 3)(x + 2) = 0$ $\qquad 4x - 3x = x$

6 $(2x - 3)(x + 2)$
$= 2x^2 + 4x - 3x - 6$
$= 2x^2 + x - 6$

7 $(2x - 3) = 0 \Rightarrow x = \dfrac{3}{2}$
$(x + 2) = 0 \Rightarrow x = -2$

Solving Quadratics

The Quadratic Formula

$$x = \frac{-b \pm \sqrt{b^2 - 4ac}}{2a}$$

Use the quadratic formula when:
- the quadratic won't factorise.
- the question mentions d.p. or s.f.
- you need exact answers or surds.

① Rearrange equation into the form $ax^2 + bx + c = 0$.

② Identify a, b and c.

③ Substitute into formula.

④ Evaluate both solutions.

Check your answers by substituting back into the original equation.

EXAMPLE

Find the solutions to $4x^2 + 3x = 5$ to 2 d.p.

① $4x^2 + 3x - 5 = 0$

The ± sign means you get two solutions.

② $a = 4$, $b = 3$, $c = -5$

③ $x = \dfrac{-3 \pm \sqrt{3^2 - 4 \times 4 \times -5}}{2 \times 4} = \dfrac{-3 \pm \sqrt{89}}{8}$

④ $x = -1.55$ (2 d.p.) or 0.80 (2 d.p.)

Completing the Square

① Multiply out initial bracket $(x + \frac{b}{2})^2$.

② Add/subtract adjusting number to match original equation.

③ Set equal to 0 and solve.

EXAMPLE

Solve $x^2 + 4x - 3 = 0$.

Check this is in the standard format first.

① $(x + 2)^2 = x^2 + 4x + 4$

② $(x + 2)^2 - 7 = x^2 + 4x + 4 - 7$
$= x^2 + 4x - 3$

Add/subtract to get −3.

③ $(x + 2)^2 - 7 = 0$
$(x + 2)^2 = 7$
$x + 2 = \pm\sqrt{7}$, so $x = -2 \pm\sqrt{7}$

... when a is not 1

① Take out a factor of 'a' from the first two terms.

② Multiply out initial bracket $a(x + \frac{b}{2a})^2$.

③ Add/subtract adjusting number to match original equation.

EXAMPLE

Write $2x^2 - 8x + 3$ in the form $a(x + m)^2 + n$.

Check this is in the standard format first.

① $2(x^2 - 4x) + 3$

Add/subtract to make this 3.

② $2(x - 2)^2 = 2x^2 - 8x + 8$

③ $2(x - 2)^2 - 5 = 2x^2 - 8x + 8 - 5$
$= 2x^2 - 8x + 3$

When a is positive, the adjusting number tells you the minimum y-value of the graph. This occurs when the brackets = 0, i.e. when $x = -m$. This also gives you the coordinates of the turning point of the graph.

Algebraic Fractions

Simplifying Algebraic Fractions

Cancel terms on the top and bottom.
Deal with one number or letter at a time.

$$\frac{8x^3y}{2x^2y^3} = \frac{{}^{4}\cancel{8} \times \cancel{x} \times \overset{x}{\cancel{x}} \times \cancel{y}}{\cancel{2} \times \cancel{x} \times \cancel{y^3}_{y^2}}$$

$\div 2$ on top and bottom
$\div x^2$ on top and bottom
$\div y$ on top and bottom

$$= \frac{4x}{y^2}$$

You might have to factorise first,
then cancel a common factor:

$$\frac{x^2 - x - 2}{x^2 + 5x + 4} = \frac{\cancel{(x+1)}(x-2)}{\cancel{(x+1)}(x+4)}$$

$$= \frac{x-2}{x+4}$$

Multiplying

Multiply tops and bottoms
of the fractions separately.

$$\frac{x}{3x+12} \times \frac{2x+8}{x-1}$$

$$= \frac{x}{3(x+4)} \times \frac{2(x+4)}{x-1}$$

$$= \frac{x \times 2}{3 \times (x-1)}$$

$$= \frac{2x}{3(x-1)}$$

Factorise and cancel first,
to make multiplying easier.

Dividing

To divide, turn the second fraction
upside down, then multiply.

$$\frac{x-5}{x^2-9} \div \frac{5x}{x-3} = \frac{x-5}{x^2-9} \times \frac{x-3}{5x}$$

Factorise using D.O.T.S.

$$= \frac{x-5}{(x-3)(x+3)} \times \frac{x-3}{5x}$$

$$= \frac{x-5}{(x+3) \times 5x}$$

$$= \frac{x-5}{5x(x+3)}$$

$$\frac{3x^2}{y^3}$$

Adding and Subtracting

① Find a common denominator.

② Multiply the top and
bottom of each fraction
by whatever gives the
common denominator.

③ Add or subtract numerators.

The common denominator
is something both
denominators divide into.

EXAMPLE

Write $\dfrac{2}{2x-1} - \dfrac{3}{x+4}$ as a
single fraction in its simplest form.

① Common denominator: $(2x-1)(x+4)$

② $\dfrac{2(x+4)}{(2x-1)(x+4)} - \dfrac{3(2x-1)}{(x+4)(2x-1)}$

③ $= \dfrac{2(x+4) - 3(2x-1)}{(2x-1)(x+4)} = \dfrac{2x+8-6x+3}{(2x-1)(x+4)}$

$$= \frac{11-4x}{(2x-1)(x+4)}$$

Collect like terms together.

Sequences

nth term of Linear Sequences

LINEAR SEQUENCES — increase/decrease by same amount each time (common difference).

1. Find the common difference — this is what you multiply n by.

2. Work out what to add/subtract.

3. Put both bits together.

EXAMPLE

Find the nth term of the sequence 7, 11, 15, 19 ...

1. $11 - 7 = 4$, $15 - 11 = 4$, etc. So common difference = 4

2. For $n = 1$, $4n = 4$. $7 - 4 = 3$, so 3 is added to each term.

3. So nth term is $4n + 3$

nth term of Quadratic Sequences

QUADRATIC SEQUENCES — have an n^2 term, so the difference between terms changes.

1. Find difference between pairs of terms.

2. Find difference between differences.

3. Divide by 2 to get coefficient of n^2.

4. Subtract n^2 term (including coefficient) from each term to get a linear sequence.

5. Find nth rule of the linear sequence.

6. Put n^2 term and linear rule together.

EXAMPLE

Find the nth term of the sequence 10, 14, 22, 34 ...

```
10   14   22   34
```

1. $+4$ $+8$ $+12$

2. $+4$ $+4$

3. $4 \div 2 = 2$, so nth term involves $2n^2$

4. term $- 2n^2$: 8, 6, 4, 2

5. Linear sequence $= -2n + 10$

6. So nth term is $2n^2 - 2n + 10$

Deciding if a Number is a Term

Set nth term rule equal to the number and solve for n. The term is in the sequence if n is an integer.

EXAMPLE

Is 37 a term in the sequence with the nth term $6n - 1$?

$6n - 1 = 37$
$6n = 38$
$n = 6.333...$

So 37 is not in the sequence.

Other Sequences

GEOMETRIC SEQUENCE — multiply/divide previous term by same number each time.

```
  ÷2  ÷2  ÷2
72  36  18  9
```

FIBONACCI-TYPE SEQUENCE — add previous two terms together.

```
      4 + 6
2   4   6   10   16
   2 + 4      6 + 10
```

Inequalities

Solving Inequalities

> means GREATER THAN   < means LESS THAN
≥ means GREATER THAN OR EQUAL TO   ≤ means LESS THAN OR EQUAL TO

Solve inequalities like equations —
but if you multiply/divide by a negative
number, flip the inequality sign.

$2x - 9 \geq 6x + 3$
$2x - 6x \geq 3 + 9$
$-4x \geq 12$ ⎯ Divided by a negative
$x \leq -3$ ⎯ number, so flip the sign.

EXAMPLE

Show $-2 \leq x < 5$ on the number line.

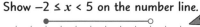

$-4\ -3\ -2\ -1\ 0\ 1\ 2\ 3\ 4\ 5\ 6\ 7$

Use ● when the value is
included, and O when it's not.

Solutions to inequalities can be given
in set notation — e.g. $\{x: -7 \leq x \leq 7\}$.

Quadratic Inequalities

If $x^2 > a^2$, then $x > a$ or $x < -a$:

$x^2 > 36$
$x^2 = 36 \Rightarrow x = -6$ or $x = 6$
So $x < -6$ or $x > 6$

If $x^2 < a^2$, then $-a < x < a$:

$2x^2 \leq 98 \Rightarrow x^2 \leq 49$ ⎯ Divide both
$x^2 = 49 \Rightarrow x = -7$ or $x = 7$ ⎯ sides by 2.
So $-7 \leq x \leq 7$

Graphical Inequalities

① Convert each inequality
to an equation.

② Draw the graph
for each equation.

Use a solid line if the inequality
uses ≤ or ≥. Use a dotted line
if the inequality uses < or >.

③ See if each inequality is true at
a specific point, to find which
side of each line you want.

④ Shade the region.

EXAMPLE

Shade the region that satisfies
$y < x + 4$, $y \leq 1 - 2x$ and $y > -1$.

① $y = x + 4$
$y = 1 - 2x$
$y = -1$

② Dotted line:
$y < x + 4$
$y > -1$
Solid line:
$y \leq 1 - 2x$

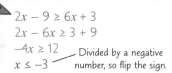

③ $y < x + 4$: $0 < 4$ is true, so
$(0, 0)$ is on **correct side** of line.

$y \leq 1 - 2x$: $0 \leq 1$ is true, so
$(0, 0)$ is on **correct side** of line.

$y > -1$: $0 > -1$ is true, so
$(0, 0)$ is on **correct side** of line.

Iterative Methods

Using Iterative Methods

ITERATIVE METHODS — repeating a calculation to get closer to the actual solution. They're used when equations are too hard to solve.

You usually keep putting the value you've just found back into the calculation.

For an equation that equals 0:

Substitute two numbers into the equation. → If the sign changes, there's a solution between the two numbers.

Decimal Search Method

EXAMPLE

The equation $x^3 - 5x - 1 = 0$ has a solution between $x = 0$ and $x = -1$. Find this solution to 1 d.p.

1. Substitute 1 d.p. values of x within the interval until the sign changes.

2. Substitute values of x with 2 d.p. until the sign changes again.

3. Repeat until values either side of the sign change are the same when rounded to required degree of accuracy.

x	$x^3 - 5x - 1$	Sign
0	-1	$-$ve
-0.1	-0.501	$-$ve
-0.2	-0.008	$-$ve
-0.3	0.473	$+$ve
-0.20	-0.008	$-$ve
-0.21	0.040739	$+$ve

3 Both -0.20 and -0.21 round to -0.2 to 1 d.p., so the solution is $x = -0.2$.

Iteration Machines

EXAMPLE

Use the iteration machine below to find a solution to $x^3 - 7x - 2 = 0$ to 1 d.p. Use the starting value $x_0 = 2$.

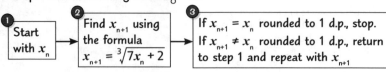

① Start with x_n → ② Find x_{n+1} using the formula $x_{n+1} = \sqrt[3]{7x_n + 2}$ → ③ If $x_{n+1} = x_n$ rounded to 1 d.p., stop. If $x_{n+1} \neq x_n$ rounded to 1 d.p., return to step 1 and repeat with x_{n+1}

Follow the instructions in the iteration machine:

$x_0 = 2$

$x_1 = 2.519... \neq x_0$ to 1 d.p.

$x_2 = 2.697... \neq x_1$ to 1 d.p.

$x_3 = 2.753... \neq x_2$ to 1 d.p.

$x_4 = 2.771... = x_3$ to 1 d.p.

x_n is the nth value, so x_{n+1} is the next value.

x_3 and x_4 both round to 2.8, so the solution is $x = 2.8$ (1 d.p.).

Simultaneous Equations

Six Steps for Easy Ones

When both equations are linear:

1. Rearrange into the form $ax + by = c$.

2. Match up the coefficients for one of the variables.

3. Add or subtract to get rid of a variable.

4. Solve the equation.

5. Substitute the value back into the original equation.

6. Check your answer works.

EXAMPLE

Solve the simultaneous equations
$5 - 2x = 3y$ and $5x + 4 = -2y$

1. $2x + 3y = 5$ (1)   Label your
   $5x + 2y = -4$ (2)   equations.

2. (1) × 5: $10x + 15y = 25$ (3)
   (2) × 2: $10x + 4y = -8$ (4)

3. (3) − (4): $0x + 11y = 33$

4. $11y = 33 \Rightarrow y = 3$

5. Sub $y = 3$ into (1): $2x + (3 \times 3) = 5$
   $\Rightarrow 2x = 5 - 9 \Rightarrow 2x = -4 \Rightarrow x = -2$

6. Sub x and y into (2):
   $(5 \times -2) + (2 \times 3) = -4$
   So the solution is $x = -2$, $y = 3$.

Seven Steps for Tricky Ones

When one equation is quadratic:

1. Rearrange one equation so a non-quadratic unknown is by itself.

2. Substitute the rearranged equation into the other equation.

3. Rearrange and solve.

4. Substitute first value into one of the equations.

5. Substitute second value into the same equation.

6. Check both pairs of solutions work.

7. Write out both pairs of solutions clearly.

EXAMPLE

Solve the simultaneous equations
$3x - y = 5$ and $3x^2 - y = 11$

1. $3x - y = 5$ (1)
   $y = 3x^2 - 11$ (2)

2. $3x - (3x^2 - 11) = 5$ (3)

3. $3x^2 - 3x - 6 = 0$
   $3(x - 2)(x + 1) = 0$
   So $x - 2 = 0 \Rightarrow x = 2$
   or $x + 1 = 0 \Rightarrow x = -1$ You'll get two values for x.

4. Sub $x = 2$ into (1):
   $6 - y = 5$, so $y = 1$

5. Sub $x = -1$ into (1):
   $-3 - y = 5$, so $y = -8$

6. Sub both x-values into (2):
   $x = 2$: $y = (3 \times 2^2) - 11 = 1$
   $x = -1$: $y = (3 \times (-1)^2) - 11 = -8$

7. $x = 2$, $y = 1$ and $x = -1$, $y = -8$

Proof

Five Facts for Algebraic Proof

1. Even Numbers — can be written as $2n$.

   'n' stands for any integer.

   It's in the pudding.

2. Odd Numbers — can be written as $2n + 1$.

3. Multiples — can be written as something $\times$ n (e.g. write multiples of 3 as $3n$).

4. Consecutive Numbers — can be written as n, n + 1, n + 2, etc.

5. The sum, difference or product of integers is always an integer.

Proof Examples

EXAMPLE

Show that the product of two odd numbers is always odd.

Odd numbers: $2a + 1$ and $2b + 1$.

$(2a + 1)(2b + 1) = 4ab + 2a + 2b + 1$
$\qquad\qquad\qquad = 2(2ab + a + b) + 1$

This can be written as $2n + 1$, where $n = 2ab + a + b$, so it must be odd.

EXAMPLE

Prove $(n - 4)^2 - (n + 1)^2 \equiv -5(2n - 3)$.

$(n - 4)^2 - (n + 1)^2$
$\equiv (n^2 - 8n + 16) - (n^2 + 2n + 1)$
$\equiv n^2 - 8n + 16 - n^2 - 2n - 1$
$\equiv -10n + 15$
$\equiv -5(2n - 3)$

The identity symbol '$\equiv$' means this is true for all values of n.

EXAMPLE

Given $\angle CED = x$, show that $\angle CAB = \frac{1}{2}(90° + x)$.

$x + 90° + \angle ECD = 180°$, so $\angle ECD = 90° - x$

$\angle ECD$ and $\angle ACB$ are vertically opposite, so $\angle ACB = 90° - x$

Triangle ABC is isosceles, so $\angle CAB = \angle ABC$

$2\angle CAB + (90° - x) = 180°$

$2\angle CAB = 90° + x$

$\angle CAB = \frac{1}{2}(90° + x)$

This is a geometric proof.

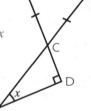

Disproof by Counter-example

Prove that a statement is false by finding a counter-example.

Keep trying numbers until you find one that doesn't work.

EXAMPLE

Disprove the statement: "The sum of two square numbers is always odd."

$1 + 4 = 5$ (odd) $4 + 9 = 13$ (odd)
$1 + 9 = 10$ (even) so the statement is false.

Functions

Evaluating Functions

FUNCTION — takes an input, processes it, outputs a value.

They're usually written like:

$$f(x) = (x + 2)^2 - 5$$

This means "take a value of x, add 2, square it, then subtract 5".

Evaluate functions by just substituting in the value of x.

$$f(-4) = (-4 + 2)^2 - 5$$
$$= (-2)^2 - 5 = -1$$

Functions can also be written like $f : x \rightarrow (x + 2)^2 - 5$.

Composite Functions

COMPOSITE FUNCTION — two functions combined into a single function.

$fg(x) \rightarrow$ put g(x) into f(x)
$gf(x) \rightarrow$ put f(x) into g(x)

Three steps for composite functions:

1. Write composite function with brackets.

2. Replace first function with its expression.

3. Substitute it into second function.

EXAMPLE

$f(x) = 4x - 1$ and $g(x) = \dfrac{3x}{2}$.

a) Find fg(x).

$$f(g(x)) = f(\tfrac{3x}{2}) = 4 \times \tfrac{3x}{2} - 1$$
$$= 6x - 1$$

b) Find gf(x).

$$g(f(x)) = g(4x - 1) = \frac{3(4x - 1)}{2}$$
$$= \frac{12x - 3}{2}$$
$$= 6x - \frac{3}{2}$$

In general, $fg(x) \neq gf(x)$.

Inverse Functions

INVERSE FUNCTION, $f^{-1}(x)$ — a function that reverses f(x).

Three steps for inverse functions:

1. Write the equation $x = f(y)$.

2. Make y the subject.

3. Replace y with $f^{-1}(x)$.

EXAMPLE

Given $f(x) = 7x - 11$, find $f^{-1}(x)$.

1. $x = 7y - 11$

   Replace f(x) with x and x with y.

2. $7y = x + 11$

   $y = \dfrac{x + 11}{7}$

3. $f^{-1}(x) = \dfrac{x + 11}{7}$

Check it reverses the function: $f(2) = 3$, and $f^{-1}(3) = 2$ ✓

Straight-Line Graphs

Straight-Line Equations

'x = a' is a **vertical**
line through 'a' on the
x-axis (e.g. x = –3)

'y = a' is a **horizontal**
line through 'a' on the
y-axis (e.g. y = –1)

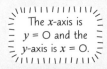
The x-axis is
$y = 0$ and the
y-axis is $x = 0$.

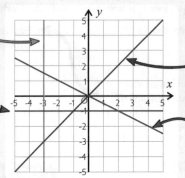

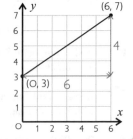
Jim took his
time to line
up the perfect
shot...

'y = x' is the
main diagonal
through the origin

'y = ax' is a **diagonal**
through the origin
with gradient 'a'
(e.g. $y = -\frac{1}{2}x$)

Equations of Straight-Line Graphs

GRADIENT — steepness of a line.

$$\text{Gradient} = \frac{\text{change in } y}{\text{change in } x}$$

1 Use any two points on the
line to find the gradient, 'm'.

2 Find the y-intercept, 'c'.

3 Write equation as y = mx + c.

EXAMPLE

1 $m = \frac{4}{6} = \frac{2}{3}$

2 $c = 3$

3 $y = \frac{2}{3}x + 3$

(graph showing line through (0, 3) and (6, 7), with 6 and 4 labelled)

Equation of a Line Through Two Points

1 Use both points to find gradient.

2 Substitute one point
into y = mx + c.

3 Rearrange to find 'c'.

4 Write equation as y = mx + c.

EXAMPLE

Find the equation of the
straight line that passes
through (–2, 12) and (4, –6).

1 $m = \frac{-6-12}{4-(-2)} = \frac{-18}{6} = -3$

2 Sub in (4, –6):
$-6 = -3(4) + c \Rightarrow -6 = -12 + c$

3 $c = -6 + 12 = 6$

4 $y = -3x + 6$

Drawing Straight-Line Graphs

'Table of 3 Values' Method

1 Draw a table with three values of x.

2 Put the x-values into the equation and work out the y-values.

3 Plot the points and draw a line through them.

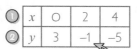 **EXAMPLE**

Draw the graph $y = -2x + 3$ for values of x from O to 4.

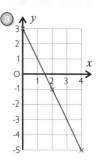

x	O	2	4
y	3	-1	-5

E.g. when $x = 2$,
$y = -2(2) + 3$
$= -4 + 3 = -1$

Using y = mx + c

1 Rearrange into the form $y = mx + c$.

2 Put a dot on the y-axis at the value of c.

3 Use m to go across and up/down an appropriate number of units. Make a dot and repeat.

4 Draw a straight line through the dots.

5 Check gradient looks correct.

EXAMPLE

Draw the graph of $3y = x + 6$.

1 $3y = x + 6 \Rightarrow y = \frac{1}{3}x + 2$

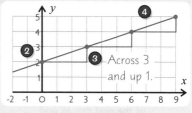

3 Across 3 and up 1.

5 A gradient of $\frac{1}{3}$ is gentle and increases from left to right. ✓

'x = 0, y = 0' Method

1 Set x = 0 and find y.

2 Set y = 0 and find x.

3 Mark and label both points. Draw a line through them.

EXAMPLE

Sketch the graph of $y = 2x - 3$.

1 When $x = O$,
$y = 2(O) - 3$
$= -3$

2 When $y = O$,
$O = 2x - 3$
$\Rightarrow x = \frac{3}{2}$

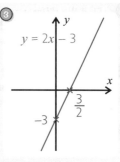

Working with Straight-Line Graphs

Parallel Line Gradients

Parallel lines have the SAME gradient
— i.e. they have the SAME m value.

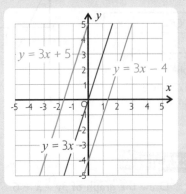

Perpendicular Line Gradients

Perpendicular lines cross at right angles.
Their gradients multiply together to give −1.

If gradient of first line = m,
then gradient of second line = $-\frac{1}{m}$.

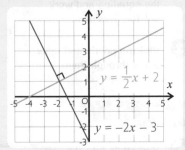

Two Steps to Find the Midpoint of a Line Segment

1. Add the x-coordinates
   of the end points and
   divide by 2.

2. Add the y-coordinates
   of the end points and
   divide by 2.

EXAMPLE

Point A has coordinates (−8, 2)
and Point B has coordinates (6, 10).
Find the coordinates of the midpoint of AB.

$$\left(\frac{-8+6}{2}, \frac{2+10}{2}\right) = \left(\frac{-2}{2}, \frac{12}{2}\right) = (-1, 6)$$

Using Ratios to Find Coordinates

1. Find the difference
   between the x-coordinates
   and the y-coordinates.

2. Use the ratio to find the
   difference between a given point
   and the point you want to find.

3. Add the differences
   to the given point.

EXAMPLE

R = (−3, −7) and S = (2, 3).
T lies on RS, so that RT:TS = 2:3.
Find the coordinates of T.

1. Difference between x-coordinates = 5
   Difference between y-coordinates = 10

2. T is $\frac{2}{2+3} = \frac{2}{5}$ along RS from R, so

   x: $\frac{2}{5} \times 5 = 2$ y: $\frac{2}{5} \times 10 = 4$

3. T = (−3 + 2, −7 + 4) = (−1, −3)

Quadratic and Cubic Graphs

Quadratic Graphs

A quadratic graph ($y = ax^2 + bx + c$) has a symmetrical bucket shape.

Three steps to plot a quadratic graph:

1 Substitute the x-values into the equation to find y-values.

2 Plot the points.

3 Join the points with a smooth curve.

The coefficient of x^2 is positive, so the curve is u-shaped.

EXAMPLE

Plot the graph of $y = x^2 + 2x - 1$.

1

x	-4	-3	-2	-1	0	1	2
y	7	2	-1	-2	-1	2	7

E.g. $y = (-4)^2 + 2(-4) - 1$
$= 16 - 8 - 1 = 7$

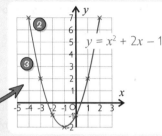

$y = x^2 + 2x - 1$

Sketching Quadratics

1 Find the x-intercepts.

2 Use symmetry to find the x-coordinate of the turning point.

3 Substitute x into the equation to find y.

4 Sketch and label graph.

The coefficient of x^2 is negative, so the curve is n-shaped.

EXAMPLE

Sketch the graph of $y = -x^2 - 2x + 3$.

1 $-x^2 - 2x + 3 = -(x + 3)(x - 1)$
So $x = -3$ and $x = 1$

2 $x = \dfrac{-3 + 1}{2} = -1$

3 $y = -(-1)^2 - 2(-1) + 3$
$= 4$

$(-1, 4)$

$(-3, 0)$ $(1, 0)$

Cubic Graphs

A cubic graph ($y = ax^3 + bx^2 + cx + d$) has a wiggle in the middle.

$+x^3$ graphs go up from bottom left: $-x^3$ graphs go down from top left:

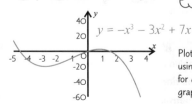

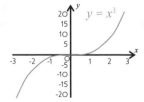

$y = x^3$

$y = -x^3 - 3x^2 + 7x$

Plot cubic graphs using the steps for quadratic graphs above.

Harder Graphs

Circle Graphs

A circle with centre (0, 0) and radius r has the equation: $$x^2 + y^2 = r^2$$

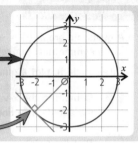

$x^2 + y^2 = 9$ is a circle with centre (0, 0).
$r^2 = 9$ so radius, r, is 3.

A radius meets a tangent at **90°**, so use perpendicular lines to find the equation of a tangent to a circle at a point.

Exponential Graphs

General form: $$y = k^x \text{ or } y = k^{-x}$$ (k is positive)

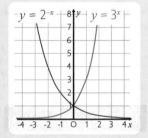

- They are always above the x-axis.
- They always go through the point (0, 1).
- If k > 1 and power is positive, graph curves upwards.
- If k is between 0 and 1 **OR** power is negative, then graph is flipped horizontally.

EXAMPLE

The graph shows how the number of bacteria (B) in a sample increases. The equation of the graph is $B = pg^h$, where h = number of hours. p and g are positive constants. Find p and g.

① Substitute in h = 0, B = 10.
$10 = pg^0 \Rightarrow 10 = p \times 1 \Rightarrow p = 10$

② Substitute in h = 2, B = 40.
$40 = 10g^2 \Rightarrow 4 = g^2 \Rightarrow g = 2$

Reciprocal Graphs

General form: $$y = \dfrac{A}{x} \text{ or } xy = A$$

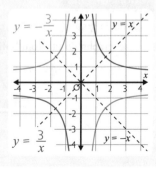

- Positive graphs in top right and bottom left quadrants.
- Negative graphs in top left and bottom right quadrants.
- Two halves of graph don't touch.
- Graphs don't exist for x = 0.
- Symmetrical about lines y = x and y = −x.

Trig Graphs and Solving Equations

Sin x and Cos x Graphs

- **Both have** y-limits **of +1 and –1.**
- **Repeat every** 360°.
- **sin graph = cos graph shifted right by 90°.**

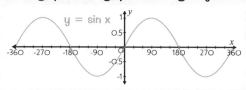

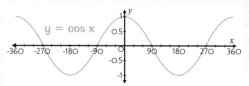

Tan x Graph

- **Goes from** –∞ **to** +∞.
- **Repeats every** 180°.
- **tan x undefined at** ±90°, ±270°, ...

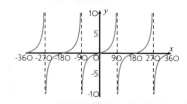

Sketch sin, cos and tan graphs by plotting important points that happen every 90°.

Solve Equations Using Graphs

EXAMPLE

By plotting the graphs, solve the simultaneous equations $y = x^2 - 5$ and $y = x - 3$.

① Draw both graphs.

② Find coordinates where graphs cross.
  $x = -1, y = -4$ and $x = 2, y = -1$

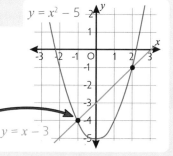

$y = x^2 - 5$

$y = x - 3$

EXAMPLE

The graph of $y = x^3 + 2x$ is shown. Find the equation of the line you'd need to draw to solve $x^3 + 3x - 1 = 0$.

① Rearrange the equation you want to solve to get the equation of the graph on its own on one side.
  $x^3 + 3x - 1 = 0$
  $x^3 + 2x = 1 - x$

② Give the equation of the line you need to draw.
  $y = 1 - x$

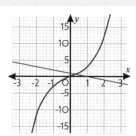

The intersection of $y = 1 - x$ and $y = x^3 + 2x$ gives the solution to $x^3 + 3x - 1 = 0$.

Graph Transformations

Translations on y-axis: $y = f(x) + a$

Adding a number to the end of the
equation translates the graph vertically.

For example:

> $y = f(x) + 2$ is a translation of 2 units UP.
>
> $y = f(x) - 4$ is a translation of 4 units DOWN.

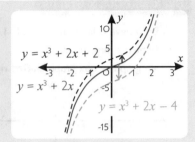

$y = x^3 + 2x + 2$

$y = x^3 + 2x$

$y = x^3 + 2x - 4$

Translations on x-axis: $y = f(x - a)$

Replacing x everywhere in the equation with
$(x - a)$ translates the graph horizontally.

Translations go the 'wrong' way:
$y = f(x - a)$ slides $y = f(x)$ 'a' units
in the positive direction (i.e. right).

For example:

> $y = f(x - 4)$ is a translation of 4 units RIGHT.
>
> $y = f(x + 3)$ is a translation of 3 units LEFT.

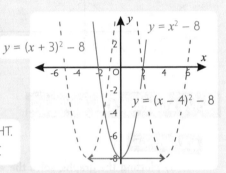

$y = x^2 - 8$

$y = (x + 3)^2 - 8$

$y = (x - 4)^2 - 8$

Reflections: $y = -f(x)$

$y = -f(x)$ is the reflection
of $y = f(x)$ in the x-axis.

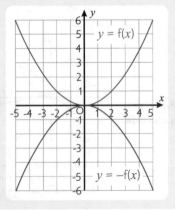

$y = f(x)$

$y = -f(x)$

Reflections: $y = f(-x)$

$y = f(-x)$ is the reflection
of $y = f(x)$ in the y-axis.

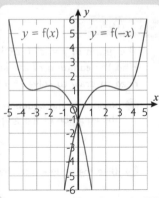

$y = f(x)$

$y = f(-x)$

Real-Life Graphs

Distance-Time Graphs

Gradient = speed

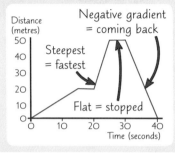

Distance (metres)

Negative gradient = coming back

Steepest = fastest

Flat = stopped

Time (seconds)

Velocity-Time Graphs

Gradient = acceleration

The units of acceleration here are m/s².

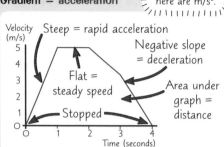

Velocity (m/s)

Steep = rapid acceleration

Negative slope = deceleration

Flat = steady speed

Stopped

Area under graph = distance

Time (seconds)

Estimate the Area Under a Velocity-Time Curve

1 Divide area into trapeziums.

2 Find area of each trapezium.

3 Add areas together to get the distance.

EXAMPLE

Estimate the distance travelled during the 15 s shown on the graph.

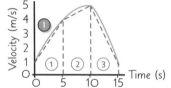

Velocity (m/s)

Time (s)

2 Area ① = 0.5 × (1 + 4) × 5 = 12.5
Area ② = 0.5 × (4 + 5) × 5 = 22.5
Area ③ = 0.5 × (5 + 1) × 5 = 15

3 12.5 + 22.5 + 15 = 50 m

Average velocity = total distance ÷ time

Gradients of Real-Life Graphs

Gradient represents rate — y-axis unit PER x-axis unit.
E.g. metres PER second (speed).

$$\text{Gradient} = \frac{\text{change in y}}{\text{change in x}}$$

Finding an Average Gradient
E.g. average speed between 2 s and 4 s:

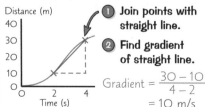

Distance (m)

Time (s)

1 Join points with straight line.

2 Find gradient of straight line.

$$\text{Gradient} = \frac{30 - 10}{4 - 2}$$
$$= 10 \text{ m/s}$$

Estimating a Gradient
E.g. speed at 3 s:

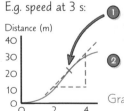

Distance (m)

Time (s)

1 Draw tangent to curve at the point.

2 Find gradient of tangent.

$$\text{Gradient} = \frac{32 - 12}{4 - 2}$$
$$= 10 \text{ m/s}$$

Section 3 — Graphs

Ratios

Writing Ratios as Fractions

Write one number on top of the other.

Or add the parts to find a fraction of the total.

My sweets   :   Your sweets

EXAMPLE

In a car park, the ratio of cars to vans is 8:3.

- There are $\frac{8}{3}$ as many cars as vans.
- There are $\frac{3}{8}$ as many vans as cars.

- There are 8 + 3 = 11 parts in total, so $\frac{8}{11}$ are cars and $\frac{3}{11}$ are vans.

Four Ways to Simplify Ratios

1 Divide all numbers by the same thing.

$$18:27 = 2:3$$
$$\div 9 \quad \div 9$$

The fraction button on your calculator can be used to help simplify ratios.

2 Multiply to get rid of fractions and decimals.

$$1.5:3.5 = 15:35 = 3:7$$
$$\times 10 \quad \div 5$$
$$\times 10 \quad \div 5$$

3 Convert to the smaller unit.

$$0.75 \text{ kg}:250 \text{ g} = 750 \text{ g}:250 \text{ g}$$
$$\div 250 \quad \div 250$$
$$= 3:1$$

4 Divide to get in the form 1:n or n:1.

$$2:5 = 1:2.5 \text{ (or } 1:\frac{5}{2}\text{)}$$
$$\div 2 \quad \div 2$$

Three Steps to Scale Up Ratios

1 Work out what one side of the ratio is multiplied by to get its actual value.

2 Multiply the other side by this number.

3 Add the two sides to find the total (if the question asks you to).

The two sides of a ratio are always in direct proportion.

EXAMPLE

A theatre audience is made up of adults and children in the ratio 3:5. There are 105 adults. How many people are there in the audience in total?

1 ×35   3:5   ×35 **2**

105:175

So there are 175 children.

3 105 + 175 = 280 people

More Ratios and Proportion

Part:Whole Ratios

PART:WHOLE RATIO — left-hand side of ratio included in right-hand side.

EXAMPLE

part:part $\qquad$ $\dfrac{part}{whole}$ $\qquad$ part:whole

$3:7 \longrightarrow \dfrac{3}{3+7} = \dfrac{3}{10} \longrightarrow 3:10$

Three Steps for Proportional Division

1. Add up the parts.

2. Divide to find one part.

3. Multiply to find the amounts.

EXAMPLE

1200 g of flour is used to make cakes, pastry and bread in the ratio 8:7:9. How much flour is used to make pastry?

1. 8 + 7 + 9 = 24 parts
2. 1 part = 1200 g ÷ 24 = 50 g
3. 7 parts = 7 × 50 g = 350 g

Two Steps for Direct Proportion

DIRECT PROPORTION — increasing one quantity increases the other proportionally.

1. Divide to find the amount for one thing.

2. Multiply to find the amount for the number of things you want.

EXAMPLE

Vivek uses 1125 ml of milk to make 5 milkshakes. How much milk will he need to make 12 milkshakes?

1. 1 milkshake uses
   1125 ml ÷ 5 = 225 ml of milk

2. 12 milkshakes will use
   225 ml × 12 = 2700 ml of milk

Two Steps for Inverse Proportion

INVERSE PROPORTION — increasing one quantity decreases the other proportionally.

1. Multiply to find the amount for one thing.

2. Divide to find the amount for the number of things you want.

EXAMPLE

Three farmers can shear 75 sheep in 45 minutes. How long would it take five farmers to shear the same number of sheep?

1. 75 sheep would take 1 farmer
   45 × 3 = 135 minutes
2. 5 farmers would take
   135 ÷ 5 = 27 minutes

Direct and Inverse Proportion

Turning Proportions into Equations

	Proportionality	Equation
y is proportional to x	$y \propto x$	$y = kx$
y is inversely proportional to x	$y \propto \dfrac{1}{x}$	$y = \dfrac{k}{x}$
y is proportional to the square of x	$y \propto x^2$	$y = kx^2$
y is inversely proportional to x cubed	$y \propto \dfrac{1}{x^3}$	$y = \dfrac{k}{x^3}$

k is a constant.

$\propto$ means 'is proportional to'.

Drawing Proportion Graphs

y is proportional to x 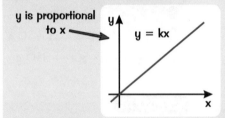 $y = kx$

y is inversely proportional to x 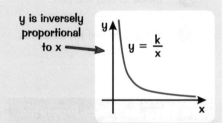 $y = \dfrac{k}{x}$

y is proportional to x² 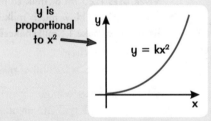 $y = kx^2$

y is inversely proportional to x³ 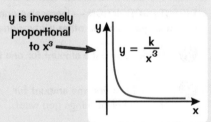 $y = \dfrac{k}{x^3}$

Four Steps for Algebraic Proportion

1. Convert **proportion** to equation.
2. Use given values to find **k**.
3. Put **k** back into equation.
4. Use equation to find value.

EXAMPLE

P is proportional to the square of Q.
When P = 320, Q = 4.
Find P when Q = 10.

1. $P \propto Q^2$, so $P = kQ^2$
2. $320 = k(4^2) = 16k$, so $k = 20$
3. $P = 20Q^2$
4. $P = 20(10^2) = 20 \times 100 = 2000$

Percentages

Three Simple Percentage Questions

1. To find a percentage of an amount, turn the percentage into a fraction/decimal then multiply.

$$35\% \text{ of } 240 = 0.35 \times 240 = 84$$

2. To find the amount after a percentage change, find the multiplier and multiply the original value by it.

EXAMPLE

Items in a sale have 12% off. What is the sale price of a hat that usually costs £7.50?

Multiplier for 12% decrease = 1 − 0.12 = 0.88
Sale price of hat = £7.50 × 0.88 = £6.60

% increase = multiplier greater than 1

% decrease = multiplier less than 1

3. To write one number as a percentage of another, divide the first by the second then multiply by 100.

$$30 \text{ as a } \% \text{ of } 250 = \frac{30}{250} \times 100 = 12\%$$

Two Steps to Find the Percentage Change

EXAMPLE

1. Find the change in amounts.

2. Use this formula:
$$\text{Percentage change} = \frac{\text{change}}{\text{original}} \times 100$$

'Change' = increase, decrease, profit, loss, etc.

A car was bought for £11 500. Four years later, it is sold for £8855. Find the percentage loss.

1. loss = £11 500 − £8855
   = £2645

2. % loss = $\frac{2645}{11\,500} \times 100$
   = 0.23 × 100 = 23%

Three Steps to Find the Original Value

EXAMPLE

1. Write the amount as a percentage of the original value.

2. Divide to find 1% of the original value.

3. Multiply by 100 to find the original value (100%).

A village has a population of 1003. The population of the village has increased by 18% since 2016. What was the population in 2016?

1. 1003 = 118%

2. 1003 ÷ 118 = 118% ÷ 118
   8.5 = 1%

3. 8.5 × 100 = 1% × 100
   850 = 100%

Working with Percentages

Simple Interest

SIMPLE INTEREST — a % of the original value is paid at regular intervals (e.g. every year). The amount of interest doesn't change.

Three steps for simple interest questions:

1 Find the % of the original value.

2 Multiply by the number of intervals.

3 Add to original value (if needed).

EXAMPLE

Lila puts £2500 in a savings account that pays 3.5% simple interest each year. How much will be in the account after 5 years?

1 3.5% of £2500
= 0.035 × £2500 = £87.50

2 5 × £87.50 = £437.50

3 £2500 + £437.50 = £2937.50

Compound Growth and Decay

Amount after n years/days/hours etc. ⟶ $N = N_0 \times (\text{multiplier})^n$ ⟵ Number of years/days/hours etc.

Initial amount   % change multiplier

EXAMPLE

A boat was bought for £15 000. It depreciates in value by 11% each year. How much will it be worth after 6 years?

N_0 = £15 000, multiplier = 1 − 0.11 = 0.89, n = 6

Value after 6 years = £15 000 × 0.89^6 = £7454.72 (to the nearest penny)

Compound Interest

COMPOUND INTEREST — a % of the new value is paid at regular intervals (e.g. every year). The amount of interest changes.

It's an example of compound growth.

EXAMPLE

Beth invests £4800 in a savings account that pays 2% compound interest per annum. How much will there be in the account after 3 years?

N_0 = £4800, multiplier = 1 + 0.02 = 1.02, n = 3

Amount after 3 years = £4800 × 1.02^3 = £5093.80 (to the nearest penny)

Measures and Units

Unit Conversions

To convert between units, multiply/divide by a conversion factor.

Metric unit conversions:

1 cm = 10 mm	1 tonne = 1000 kg
1 m = 100 cm	1 litre = 1000 ml
1 km = 1000 m	1 litre = 1000 cm³
1 kg = 1000 g	1 cm³ = 1 ml

For metric-imperial conversions, conversion factors will be given.

EXAMPLE

Use the conversion 5 miles ≈ 8 km to work out how many metres there are in 13 miles.

To convert from miles to km, divide by 5 then multiply by 8:

13 miles ≈ 13 ÷ 5 × 8 = 2.6 × 8
= 20.8 km

Then convert km to m using the conversion factor 1000:

20.8 km = 20.8 × 1000 = 20 800 m

Converting Areas

1 m² = 100 cm × 100 cm
= 10 000 cm²

1 cm² = 10 mm × 10 mm
= 100 mm²

Converting Volumes

1 m³ = 100 cm × 100 cm × 100 cm
= 1 000 000 cm³

1 cm³ = 10 mm × 10 mm × 10 mm
= 1000 mm³

Speed, Density and Pressure

$$\text{SPEED} = \frac{\text{DISTANCE}}{\text{TIME}}$$

Units of speed: distance travelled per unit time, e.g. km/h, m/s

$$\text{DENSITY} = \frac{\text{MASS}}{\text{VOLUME}}$$

Units of density: mass per unit volume, e.g. kg/m³, g/cm³

$$\text{PRESSURE} = \frac{\text{FORCE}}{\text{AREA}}$$

Units of pressure: force per unit area, e.g. N/m² (or pascals)

Use formula triangles to rearrange formulas. Cover up the thing you want and write down what's left.

EXAMPLE

The density of copper is 8.96 g/cm³. What is the mass of a copper cube with volume 0.008 m³?

Convert volume to cm³:
0.008 m³ × 100 × 100 × 100 = 8000 cm³

Use the formula triangle to get the formula for mass:
mass = density × volume
= 8.96 × 8000
= 71 680 g

Section 4 — Ratio, Proportion and Rates of Change

Geometry

Five Angle Rules

1 Angles in a triangle add up to 180°.

2 Angles on a straight line add up to 180°.

3 Angles in a quadrilateral add up to 360°.

4 Angles round a point add up to 360°.

5 Isosceles triangles have 2 identical sides and 2 identical angles.

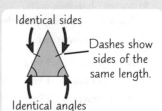

Identical sides

Dashes show sides of the same length.

Identical angles

Angles Around Parallel Lines

When a line crosses two parallel lines:

- Two bunches of angles are formed.
- There are only two different angles (a and b).
- Vertically opposite angles are equal.

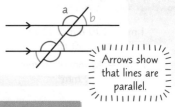

Arrows show that lines are parallel.

Alternate Angles

Found in a Z-shape:

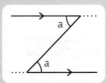

Alternate angles are the same.

Corresponding Angles

Found in an F-shape:

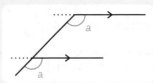

Corresponding angles are the same.

Allied Angles

Found in a C- or U-shape:

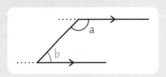

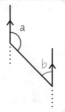

Allied angles add up to 180°.

$$a + b = 180°$$

Polygons

Regular Polygons

Name	Pentagon	Hexagon	Heptagon	Octagon	Nonagon	Decagon
No. of sides	5	6	7	8	9	10

Number of lines of symmetry = Number of sides = Order of rotational symmetry

Interior and Exterior Angles

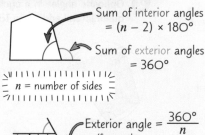

Sum of interior angles
= $(n - 2) \times 180°$

Sum of exterior angles
= $360°$

n = number of sides

Exterior angle = $\dfrac{360°}{n}$
(for regular polygons only)

Interior angle
= $180°$ − exterior angle

Four Types of Triangles

EQUILATERAL

$60°$

- 3 lines of symmetry
- Rotational sym. order 3

ISOSCELES

- 1 line of symmetry
- No rotational symmetry

RIGHT-ANGLED
$90°$

No symmetry
unless isosceles

SCALENE

No rotational symmetry = order 1

Six Types of Quadrilaterals

SQUARE

- 4 lines of symmetry
- Rotational sym. order 4
- Diagonals equal length, cross at right angles

RECTANGLE

- 2 lines of symmetry
- Rotational sym. order 2
- Diagonals equal length

RHOMBUS
Add up to $180°$

- 2 lines of symmetry
- Rotational sym. order 2
- Diagonals cross at right angles

PARALLELOGRAM
Add up to $180°$

- No lines of symmetry
- Rotational sym. order 2

TRAPEZIUM

- No lines of symmetry (unless isosceles)
- No rotational symmetry

KITE

- 1 line of symmetry
- No rotational symmetry
- Diagonals cross at right angles

Circle Geometry

Three Rules with Tangents

1 A tangent and a radius meet at 90°.

2 Tangents from the same point are the same length.

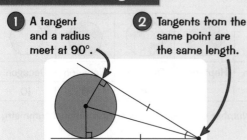

3 ALTERNATE SEGMENT THEOREM

The angle between a tangent and a chord is equal to the angle in the opposite segment.

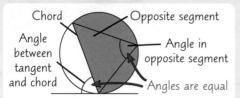

Chord — Opposite segment

Angle between tangent and chord — Angle in opposite segment

Angles are equal

Two Rules with Polygons

1 Two radii form an isosceles triangle.

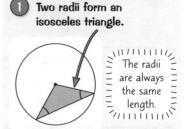

The radii are always the same length.

2 Opposite angles in a cyclic quadrilateral add up to 180°.

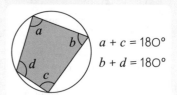

$a + c = 180°$

$b + d = 180°$

Four More Rules

1 The perpendicular bisector of a chord passes through the centre.

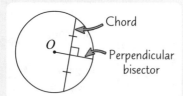

Chord

Perpendicular bisector

2 Angle made at the centre is twice the angle made at the circumference.

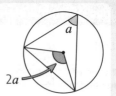

3 Angle in a semicircle is 90°.

Two angles in opposite segments add up to 180°.

4 Angles in the same segment are equal.

$a + b = 180°$

Congruent and Similar Shapes

Four Conditions for Congruent Triangles

CONGRUENT — same size and same shape.

Shapes are congruent under translation, rotation and reflection.

Condition	❶ SSS	❷ ASA	❸ SAS	❹ RHS
Description	three sides the same	two angles and corresponding side match up	two sides and angle between them match up	right angle, hypotenuse and another side all match up
Diagrams				

Two Steps to Prove Congruence

① Write down everything you know.

② State which condition holds and why.

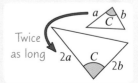

EXAMPLE

O is the centre of this circle. Prove that triangles *ABO* and *CDO* are congruent.

① *AO*, *BO*, *CO* and *DO* are all radii, so they're equal.

Angles *AOB* and *COD* are vertically opposite, so they're equal.

② SAS — two sides and the angle between them match up, so *ABO* and *CDO* are congruent.

Three Conditions for Similar Triangles

Shapes are similar under enlargement.

SIMILAR — same shape, different size.

① All angles match up.

② All sides are proportional.

③ Two sides proportional and angle between is the same.

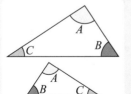

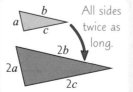
All sides twice as long.

Twice as long

The Four Transformations

Translation

Amount a shape moves is given by $\begin{pmatrix} x \\ y \end{pmatrix}$.

x = horizontal movement
y = vertical movement

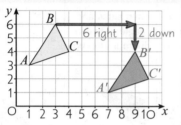

Translation from
ABC to A'B'C': $\begin{pmatrix} 6 \\ -2 \end{pmatrix}$

Rotation

To describe a rotation you need:

1 the angle   **2** the direction

3 the centre of rotation

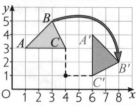

Rotation from ABC to A'B'C':
90° clockwise about (4, 1)
1   **2**   **3**

Reflection

Describe by giving the
equation of the mirror line.

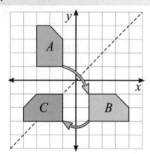

B is a reflection of A in y = x
C is a reflection of B in the y-axis

Enlargement

To describe an enlargement you need:

1 the scale factor = $\dfrac{\text{new length}}{\text{old length}}$

2 the centre of enlargement

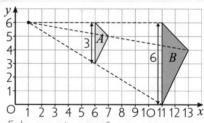

Enlargement
from A to B: **1** $\dfrac{6}{3} = 2$ **2** (1, 6)

Four Facts about Scale Factors

1 If bigger than 1, shape gets bigger.

2 If smaller than 1, shape gets smaller.

3 If negative, shape goes to other
side of centre of enlargement.
Scale factor of –1 = rotation of 180°.

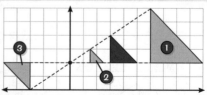

4 They give relative distance of new and old points from centre of enlargement.

Perimeter and Area

Triangles and Quadrilaterals

Area of rectangle = **length × width**

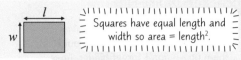
Squares have equal length and width so area = length².

Area of triangle = $\frac{1}{2}$ **× base × vertical height**

Area of parallelogram = **base × vertical height**

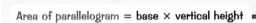

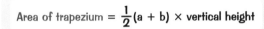

Area of trapezium = $\frac{1}{2}$**(a + b) × vertical height**

Split composite shapes into triangles and quadrilaterals.
Work out each area and add together.
Only include outside edges when adding up perimeters.

5 cm²
10 cm²
Total area: 15 cm²

Circles

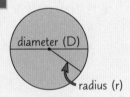
diameter (D)
radius (r)

Area = **π × (radius)²**
     = **πr²**

Circumference = **π × diameter = 2 × π × radius**
              = **πD = 2πr**

Arcs and Sectors

Major Sector
Minor Sector
Major Arc
Minor Arc
x

Area of sector = $\frac{x}{360}$ **× area of full circle**

Length of arc = $\frac{x}{360}$ **× circumference of full circle**

Segments

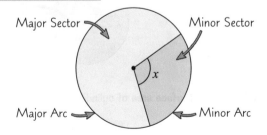
Major Segment
Chord
Minor Segment

Area of minor segment
= **area of minor sector**
− area of triangle

3D Shapes and Surface Area

Parts of 3D Shapes

If you're asked to find the number of vertices/edges/faces, just count them up — don't forget hidden ones.

E.g. this cube has 8 vertices, 12 edges and 6 faces.

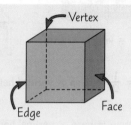
Vertex

Edge

Face

Janus approves of this page.

Three Projections

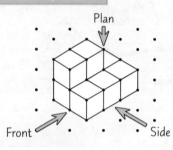
Plan

Front

Side

1 **Front elevation**

2 **Side elevation**

3 **Plan**

Dotty paper is called isometric paper.

Surface Area

SURFACE AREA — total area of all faces.

Surface area of solid = area of net

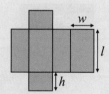
w

l

h

Surface area of sphere = $4\pi r^2$

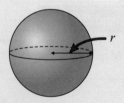
r

Surface area of cone = $\pi rl + \pi r^2$

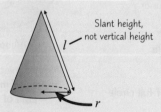
Slant height, not vertical height

l

r

Surface area of cylinder = $2\pi rh + 2\pi r^2$

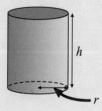
h

r

Section 5 — Geometry and Measures

Volume and Enlargement

Six Volume Formulas

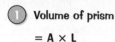

1 Volume of prism

$$= A \times L$$

A = constant area of cross-section

2 Volume of cylinder

$$= \pi r^2 h$$

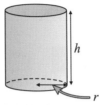

3 Volume of sphere

$$= \frac{4}{3}\pi r^3$$

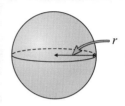

4 Volume of pyramid

$$= \frac{1}{3} \times \text{base area} \times h_v$$

base area

5 Volume of cone

$$= \frac{1}{3}\pi r^2 h_v$$

6 Volume of frustum

$$= \frac{1}{3}\pi R^2 H - \frac{1}{3}\pi r^2 h$$

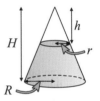

Rates of Flow

RATE OF FLOW — how fast volume is changing.

EXAMPLE

A cylinder with radius 10 cm and height 8 cm is filled with water at 1 litre per minute. How long does this take to the nearest second?

Find total volume:

$V = \pi \times 10^2 \times 8 = 2513.2...$ cm^3

Convert units: ⟋1 L = 1000 cm^3

1000 cm^3/min = 16.6... cm^3/s

Solve for time:

2513.2... ÷ 16.6... = 151 s (to nearest s)

Enlargement of Areas and Volumes

If a shape changes by a scale factor of n:

Sides are n times bigger (1:n).

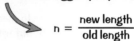
$$n = \frac{\text{new length}}{\text{old length}}$$

Areas are n^2 times bigger (1:n^2).

$$n^2 = \frac{\text{new area}}{\text{old area}}$$

Volumes are n^3 times bigger (1:n^3).

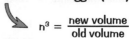
$$n^3 = \frac{\text{new volume}}{\text{old volume}}$$

Triangle Construction

Constructing Triangles

There's only one triangle you can draw if you're given:

 SSS   **SAS**   **ASA**   **RHS**

> There are two triangles you could draw if you know two sides and an angle that isn't between them.

Four Steps for Three Known Sides

EXAMPLE

1. Roughly sketch and label the triangle.

2. Accurately draw and label the base line.

3. Set compasses to each side length, then draw an arc at each end.

4. Join up the ends with the arc intersection. Label points and sides.

Construct triangle ABC where AB = 3 cm, BC = 2 cm, AC = 2.5 cm.

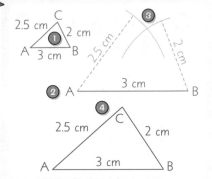

Five Steps for Known Sides and Angles

EXAMPLE

1. Roughly sketch and label the triangle.

2. Accurately draw and label the base line.

3. Use a protractor to measure the angles and mark out with dots.

4. **ASA** Draw lines from ends through dots. Label the intersection.

   SAS Measure towards dot. Label the point.

   RHS Draw line from one end through dot. Draw arc from other end. Label the intersection.

5. Join up the points. Label known sides and angles.

Construct triangle XYZ where XY = 2 cm, angle YXZ = 70°, angle XYZ = 40°.

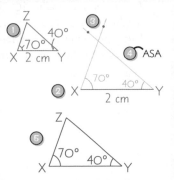

Loci and Construction

Four Different Types of Loci

LOCI — lines or regions showing all points that fit a given rule.

1 Locus of points at a fixed distance from a given point:

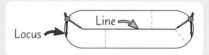

2 Locus of points at a fixed distance from a given line:

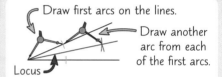

3 Locus of points equidistant from two given lines:

░ This locus bisects the angle between the two lines. ░

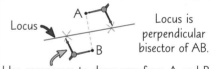

Draw first arcs on the lines.

Draw another arc from each of the first arcs.

Locus

4 Locus of points equidistant from two given points:

░ When constructing any of these four loci, keep your compass settings the same. ░

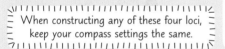

Locus is perpendicular bisector of AB.

Use compasses to draw arcs from A and B.

Constructing 60° Angles

Keep compass settings the same for 60° angles.

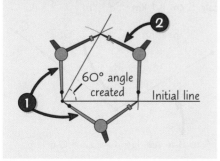

60° angle created

Initial line

Constructing 90° Angles

2 Increase compass setting for this step.

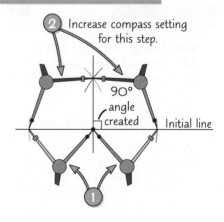

90° angle created

Initial line

Construction and Bearings

Drawing the Perpendicular From a Point to a Line

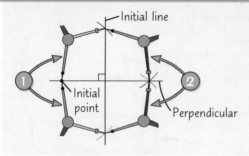

Initial line

Initial point

Perpendicular

Constructing a line that is perpendicular to the one you've just drawn gives a line that is parallel to the initial line.

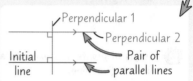

Perpendicular 1

Perpendicular 2

Initial line

Pair of parallel lines

Three Steps to Find Bearings

1 **Put your pencil at the point you're going from.**

2 **Draw a north line at that point.**

3 **Measure the angle clockwise from the north line to the line joining the two points.**

Bearings must be given as 3 figures — e.g. 090° rather than 90°.

EXAMPLE

Find the bearing of X from Y.

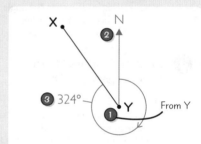

X

N

324°

Y

From Y

So the bearing of X from Y is 324°.

Scale Drawings

Scale drawings and maps show the positions of objects and the distances between them.

Real-life distance
= map distance × scale factor

Be wary of units. They're not usually the same for real life and the map.

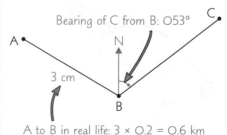

Scale: 5 cm = 1 km. So 1 cm = 0.2 km.

Bearing of C from B: 053°

C

A

N

3 cm

B

A to B in real life: 3 × 0.2 = 0.6 km

Pythagoras' Theorem

Pythagoras' Theorem

Uses two sides to find third side:

$$a^2 + b^2 = c^2$$

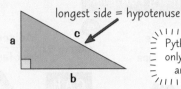

longest side = hypotenuse

Pythagoras' theorem only works for right-angled triangles.

Find a Missing Length

1. Write down formula.

2. Put in numbers.

3. Rearrange equation.

4. Take square root.

5. Give answer in correct form.

EXAMPLE

Find the length of AB to 1 d.p.

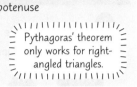

9 m   4 m

1. $a^2 + b^2 = c^2$

2. $AB^2 + 4^2 = 9^2$ c = AC (the longest side)

3. $AB^2 = 9^2 - 4^2 = 81 - 16 = 65$

4. $AB = \sqrt{65} = 8.062...$ m

5. 8.1 m (1 d.p.)

Find Distance Between Points

1. Sketch triangle.

2. Subtract coordinates to find shorter lengths.

3. Use Pythagoras to find hypotenuse.

4. Give answer in correct form.

EXAMPLE

Point L has coordinates (−1, 0).
Point M has coordinates (3, −2).
Find the exact distance LM.

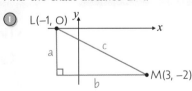

1. L(−1, 0)   M(3, −2)

2. Length of side a = 0 − −2 = 2
   Length of side b = 3 − −1 = 4

3. $a^2 + b^2 = c^2$
   $2^2 + 4^2 = c^2$
   $c^2 = 4 + 16 = 20$

4. $c = \sqrt{20} = 2\sqrt{5}$

The distance is the hypotenuse, so you don't need to rearrange the equation.

'Exact' means leave it in surd form (simplified if possible).

Trigonometry

Three Trigonometry Formulas

1 $\text{Sin } x = \dfrac{\text{Opposite}}{\text{Hypotenuse}}$

SOH $\dfrac{O}{S \times H}$

2 $\text{Cos } x = \dfrac{\text{Adjacent}}{\text{Hypotenuse}}$

CAH $\dfrac{A}{C \times H}$

3 $\text{Tan } x = \dfrac{\text{Opposite}}{\text{Adjacent}}$

TOA $\dfrac{O}{T \times A}$

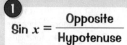

Opposite (O) — side opposite angle x

Adjacent (A) — side next to angle x

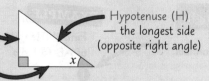

Hypotenuse (H) — the longest side (opposite right angle)

These formulas only work on right-angled triangles.

Find a Missing Length

1 Label sides O, A and H.

2 Choose formula.

3 Use a formula triangle to rearrange formula.

4 Put in numbers and work out length.

EXAMPLE

Find the length of g to 2 s.f.

1 10 cm H 55° g A O

2 A and H are involved, so use CAH.

SOH (CAH) TOA

3 $\dfrac{A}{C \times H}$ $A = C \times H$ — You're finding A.

4 $g = \cos 55° \times 10 = 5.735... = 5.7$ cm (2 s.f.)

Find a Missing Angle

1 Label sides O, A and H.

2 Choose formula.

3 Use a formula triangle to rearrange formula.

4 Put in numbers.

5 Take inverse to find angle.

EXAMPLE

Find angle x to 1 d.p.

1 12 m O 4 m H x A

2 O and A are involved, so use TOA.

SOH CAH (TOA)

3 $\dfrac{O}{T \times A}$ $T = \dfrac{O}{A}$ — Cover T to find formula.

4 $\tan x = \dfrac{12}{4} = 3$

5 $x = \tan^{-1}(3) = 71.565...° = 71.6°$ (1 d.p.)

Common Trig Values and the Sine Rule

Common Trig Values

These values can be worked out from these two triangles.

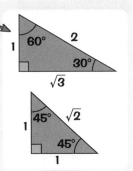

$$\sin 30° = \frac{1}{2} \qquad \sin 60° = \frac{\sqrt{3}}{2} \qquad \sin 45° = \frac{1}{\sqrt{2}}$$

$$\cos 30° = \frac{\sqrt{3}}{2} \qquad \cos 60° = \frac{1}{2} \qquad \cos 45° = \frac{1}{\sqrt{2}}$$

$$\tan 30° = \frac{1}{\sqrt{3}} \qquad \tan 60° = \sqrt{3} \qquad \tan 45° = 1$$

These values cannot be worked out using triangles.

$$\sin 0° = 0 \qquad \sin 90° = 1$$

$$\cos 0° = 1 \qquad \cos 90° = 0 \qquad \tan 0° = 0$$

Use common trig values to find exact values in triangles.

The Sine Rule

$$\frac{a}{\sin A} = \frac{b}{\sin B} = \frac{c}{\sin C}$$

You only use two bits of the formula at a time. You can turn the formula upside down if you're finding an angle.

Use when given:

2 ANGLES + ANY SIDE

EXAMPLE

Find the length AC.

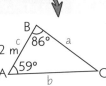

① Find missing angle.
$$C = 180° - 86° - 59° = 35°$$

② Put numbers in sine rule.
$$\frac{b}{\sin B} = \frac{c}{\sin C} \implies \frac{AC}{\sin 86°} = \frac{2}{\sin 35°}$$

③ Rearrange to find length.
$$AC = \frac{2 \times \sin 86°}{\sin 35°} = 3.5 \text{ m (2 s.f.)}$$

2 SIDES + ANGLE NOT ENCLOSED BY THEM

EXAMPLE

Find angle A.

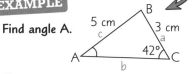

① Put numbers in sine rule.
$$\frac{a}{\sin A} = \frac{c}{\sin C} \implies \frac{3}{\sin A} = \frac{5}{\sin 42°}$$

② Rearrange to find sin A.
$$\sin A = \frac{3 \times \sin 42°}{5} = 0.401...$$

③ Take inverse to find angle.
$$A = \sin^{-1}(0.401...) = 23.7° \text{ (1 d.p.)}$$

The Cosine Rule and Area of a Triangle

The Cosine Rule

To find a side: $a^2 = b^2 + c^2 - 2bc \cos A$

To find an angle: $\cos A = \dfrac{b^2 + c^2 - a^2}{2bc}$

Use when given:

2 SIDES + ANGLE ENCLOSED BY THEM

ALL 3 SIDES, NO ANGLES

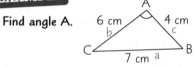

EXAMPLE

Find the length BC.

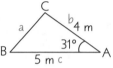

❶ Put numbers in cosine rule.
$a^2 = b^2 + c^2 - 2bc \cos A$
$= 4^2 + 5^2 - 2 \times 4 \times 5 \cos 31°$
$= 6.7133...$

❷ Take square root to find length.
$a = \sqrt{6.7133...} = 2.6$ m (2 s.f.)

EXAMPLE

Find angle A.

❶ Put numbers in cosine rule.
$\cos A = \dfrac{b^2 + c^2 - a^2}{2bc} = \dfrac{6^2 + 4^2 - 7^2}{2 \times 6 \times 4}$
$= 0.0625$

❷ Take inverse to find angle.
$A = \cos^{-1}(0.0625) = 86.4°$ (1 d.p.)

Area of Triangle

$$\text{Area of triangle} = \frac{1}{2} ab \sin C$$

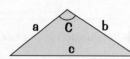

Use when given two sides and the angle enclosed by them.

Two steps to find the area of a triangle:

① Label the sides and angle.

② Put numbers in formula.

EXAMPLE

Find the area of triangle PQR.

❶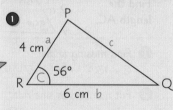

❷ Area $= \dfrac{1}{2} ab \sin C$
$= \dfrac{1}{2} \times 4 \times 6 \times \sin 56°$
$= 9.9$ cm^2 (2 s.f.)

3D Pythagoras and Trigonometry

3D Pythagoras

To find the length of a diagonal of a cuboid:

$$a^2 + b^2 + c^2 = d^2$$

Two steps for other 3D shapes:

1. Form a cuboid that has diagonal d within the 3D shape.

2. Put numbers into formula.

EXAMPLE

Find the exact length of BD. The vertical height of the prism is 4 m.

1.

2. $a^2 + b^2 + c^2 = d^2$
   $5^2 + 3^2 + 4^2 = BD^2$
   $BD = \sqrt{50} = 5\sqrt{2}$ m

Angle Between Line and Plane

1. Draw right-angled triangle between the line and plane.

2. Sketch triangle in 2D.

3. Use Pythagoras to find any missing sides.

   You need to know two sides.

4. Use trig to find angle.

EXAMPLE

Find the angle that the diagonal BH makes with the cuboid's base.

1.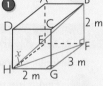

2.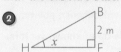

Use triangle FGH on cuboid's base.

3. $FH^2 = 2^2 + 3^2 = 13$, so $FH = \sqrt{13}$

4. $T = \dfrac{O}{A} \Rightarrow \tan x = \dfrac{2}{\sqrt{13}} = 0.5547...$
   $\Rightarrow x = \tan^{-1}(0.5547...) = 29.0°$ (1 d.p.)

Find a Length Using an Angle

1. Draw right-angled triangle containing angle and missing length.

2. Sketch triangle in 2D.

3. Use Pythagoras to find a side.

4. Use trig to find length.

EXAMPLE

In this square-based pyramid, angle ACE = 50°. Find length AC.

1.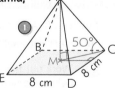

2.

Use triangle on base with shorter sides of length 4 cm (as M is the midpoint).

3. $MC^2 = 4^2 + 4^2 = 32$, so $MC = \sqrt{32}$ cm

4. $C = \dfrac{A}{H} \Rightarrow \cos 50° = \dfrac{\sqrt{32}}{AC}$
   $AC = \sqrt{32} \div \cos 50° = 8.8$ cm (2 s.f.)

Vectors

Vector Notation and Ratios

This vector can be written as $\overrightarrow{CD}$, **b**, <u>b</u> or $\binom{4}{-1}$.

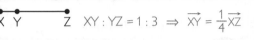

Ratios can show relative lengths of sections on a line.

If you know one vector, you can use it to find others.

$XY : YZ = 1 : 3 \Rightarrow \overrightarrow{XY} = \frac{1}{4}\overrightarrow{XZ}$

Multiplying a Vector by a Scalar

Scalar multiples are parallel.

Multiplying a vector by:

+ a positive number changes its size only.

− a negative number reverses the direction too.

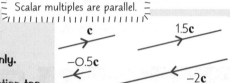

Adding and Subtracting Vectors

To describe a movement between points:

1 Find route made up of known vectors.

2 Add vectors along route. Subtract vectors travelled in reverse direction.

For column vectors: add/subtract top numbers, then bottom numbers.

E.g. $\binom{4}{-1} - \binom{2}{3} = \binom{2}{-4}$

EXAMPLE

M is the midpoint of AB. Find vector $\overrightarrow{CA}$.

1 $\overrightarrow{AM} = \underline{b}$ as M is the midpoint.
So $\overrightarrow{CA} = \overrightarrow{CB} + \overrightarrow{BM} + \overrightarrow{MA}$.

2 $= \underline{c} - \underline{b} - \underline{b} = \underline{c} - 2\underline{b}$

You're going backwards along <u>b</u>, so subtract.

Showing Points are on a Straight Line

Points A, B, C lie on a straight line if $\overrightarrow{AB}$ is a scalar multiple of $\overrightarrow{BC}$ or $\overrightarrow{AC}$.

1 Work out the vectors between points on the line.

2 Check vectors are scalar multiples of each other.

3 Explain your reasoning.

EXAMPLE

Show that ABC is a straight line.

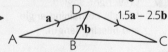

1 Find $\overrightarrow{AB}$ and $\overrightarrow{BC}$.
$\overrightarrow{AB} = \underline{a} - \underline{b}$
$\overrightarrow{BC} = \underline{b} + 1.5\underline{a} - 2.5\underline{b} = 1.5\underline{a} - 1.5\underline{b}$

2 $\overrightarrow{BC} = 1.5(\underline{a} - \underline{b})$, so $\overrightarrow{BC} = 1.5\overrightarrow{AB}$

3 $\overrightarrow{BC}$ is a scalar multiple of $\overrightarrow{AB}$, so ABC is a straight line.

Probability Basics

The Probability Scale

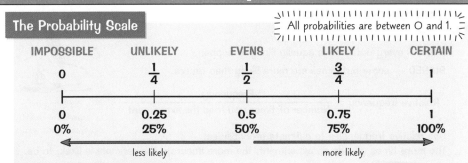

All probabilities are between O and 1.

IMPOSSIBLE	UNLIKELY	EVENS	LIKELY	CERTAIN
0	$\frac{1}{4}$	$\frac{1}{2}$	$\frac{3}{4}$	1
0	0.25	0.5	0.75	1
0%	25%	50%	75%	100%

less likely   more likely

The Probability Formula

$$\text{Probability} = \frac{\text{Number of ways for something to happen}}{\text{Total number of possible outcomes}}$$

You can only use this formula if all the outcomes are equally likely.

EXAMPLE

What is the probability of picking a prime number at random from a bag of counters numbered 1-15?

The prime numbers between 1 and 15 are 2, 3 5, 7, 11 and 13 — 6 in total.

$$\text{Probability} = \frac{\text{number of ways of picking a prime}}{\text{total number of possible outcomes}} = \frac{6}{15} = \frac{2}{5}$$

There are 15 counters so 15 possible outcomes.

Probabilities of Events

If only one possible outcome can happen at a time, the probabilities of all possible outcomes add up to 1. As events either happen or don't:

P(event happens) + P(event doesn't happen) = 1

So:

P(event doesn't happen) = 1 – P(event happens)

EXAMPLE

The probability of getting a 5 on a spinner is O.65. What is the probability of not getting a 5?

P(not 5) = 1 – P(5)
= 1 – O.65 = O.35

Sample Space Diagrams

These show all possible outcomes.

E.g. All possible outcomes when two spinners numbered 1, 2, 3 and 2, 4, 6 are spun and the results multiplied.

×	1	2	3
2	2	4	6
4	4	8	12
6	6	12	18

The Product Rule

Number of ways to carry out a combination of activities = number of ways to carry out each activity multiplied together

Number of ways to roll 3 fair 6-sided dice = 6 × 6 × 6 = 216

Probability Experiments

Repeating Experiments

FAIR — every outcome is equally likely to happen.

BIASED — some outcomes are more likely than others.

$$\text{Relative frequency} = \frac{\text{Frequency}}{\text{Number of times you tried the experiment}}$$

Repeating the experiment hadn't improved Robin's accuracy.

Use relative frequencies to estimate probabilities.

The more times you do an experiment, the more accurate the estimate is likely to be.

EXAMPLE

A spinner labelled A to D is spun 100 times. It lands on C 48 times. Find the relative frequency of spinning a C and say whether you think this spinner is biased.

Relative frequency of C = $\frac{48}{100}$ = 0.48

If the spinner was fair, you'd expect the relative frequency of C to be 1 ÷ 4 = 0.25. 0.48 is much larger than 0.25, so the spinner is probably biased.

Frequency Trees

Used to record results when experiments have more than one step. For example:

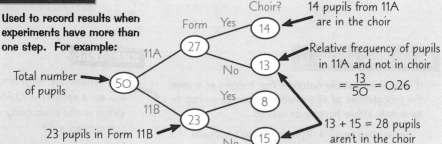

Total number of pupils

Choir?

14 pupils from 11A are in the choir

Relative frequency of pupils in 11A and not in choir = $\frac{13}{50}$ = 0.26

23 pupils in Form 11B

13 + 15 = 28 pupils aren't in the choir

Expected Frequency

EXPECTED FREQUENCY — how many times you'd expect something to happen in a certain number of trials.

Expected frequency
   = probability × number of trials

Use the relative frequency from previous experiments if you don't know the probability.

EXAMPLE

A fair 6-sided dice is rolled 360 times. How many times would you expect it to land on 4?

P(4) = $\frac{1}{6}$

Expected frequency of 4 = $\frac{1}{6}$ × 360

   = 60

The AND/OR Rules

The AND Rule for Independent Events

INDEPENDENT EVENTS — where one event happening doesn't affect the probability of another event happening.

If you select a second item after replacing the first, the events are independent.

For independent events A and B:

P(A and B) = P(A) × P(B)

EXAMPLE

A fair dice is rolled and a fair coin is tossed. What is the probability of rolling a 2 and getting heads?

$P(2) = \frac{1}{6}$ and $P(heads) = \frac{1}{2}$

Rolling a dice and tossing a coin are independent, so:

$P(2 \text{ and heads}) = \frac{1}{6} \times \frac{1}{2} = \frac{1}{12}$

The OR Rule

For any events A and B:

$P(A \text{ or } B) = P(A) + P(B) - P(A \text{ and } B)$

If A and B are mutually exclusive, P(A and B) = 0. So the OR rule becomes:

$P(A \text{ or } B) = P(A) + P(B)$

Mutually exclusive events can't happen together.

EXAMPLE

A fair dice is rolled and a fair coin is tossed. What is the probability of rolling a 2 or getting heads?

From above, $P(2) = \frac{1}{6}$, $P(heads) = \frac{1}{2}$ and $P(2 \text{ and heads}) = \frac{1}{12}$

So $P(2 \text{ or heads}) = \frac{1}{6} + \frac{1}{2} - \frac{1}{12}$
$= \frac{7}{12}$

Conditional Probability

DEPENDENT EVENTS — where one event happening affects the probability of another event happening.

If you select a second item without replacing the first, the events are dependent.

CONDITIONAL PROBABILITY OF A GIVEN B — the probability of event A happening given that event B happens.

For dependent events A and B, the AND rule is:

$P(A \text{ and } B) = P(A) \times P(B \text{ given } A)$

P(B given A) can be written P(B|A).

EXAMPLE

The probability that Abi has pasta for tea is 0.6.
The probability that Abi has yoghurt for pudding given that she has pasta for tea is 0.7. What is the probability that Abi has pasta for tea and yoghurt for pudding?

P(pasta and yoghurt)
= P(pasta) × P(yoghurt given pasta)
= 0.6 × 0.7 = 0.42

If A and B are independent, then P(A given B) = P(A) and P(B given A) = P(B).

Tree and Venn Diagrams

Tree Diagrams

Used to work out probabilities for combinations of events — e.g. for a bag containing 3 red and 2 blue counters that are selected at random without replacement:

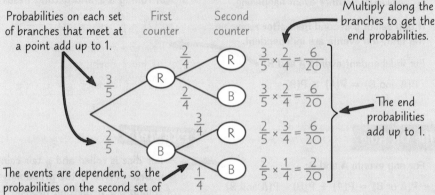

Probabilities on each set of branches that meet at a point add up to 1.

First counter

Second counter

Multiply along the branches to get the end probabilities.

$$\frac{3}{5} \times \frac{2}{4} = \frac{6}{20}$$

$$\frac{3}{5} \times \frac{2}{4} = \frac{6}{20}$$

$$\frac{2}{5} \times \frac{3}{4} = \frac{6}{20}$$

$$\frac{2}{5} \times \frac{1}{4} = \frac{2}{20}$$

The end probabilities add up to 1.

The events are dependent, so the probabilities on the second set of branches change depending on the outcomes of the first event.

If the counters were replaced, the events would be independent so the probabilities on each set of branches would be the same.

Add up the end probabilities to answer questions:

E.g. P(one red, one blue) = P(R, B) + P(B, R)

$$= \frac{6}{20} + \frac{6}{20} = \frac{12}{20} = \frac{3}{5}$$

Sets and Venn Diagrams

SET — a collection of elements (e.g. numbers), written in curly brackets {}.

VENN DIAGRAM — a diagram where sets are represented by overlapping circles.

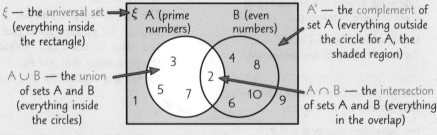

ξ — the universal set (everything inside the rectangle)

A (prime numbers)   B (even numbers)

A' — the complement of set A (everything outside the circle for A, the shaded region)

$A \cup B$ — the union of sets A and B (everything inside the circles)

$A \cap B$ — the intersection of sets A and B (everything in the overlap)

n(A) — the number of elements in set A. So here, n(A) = 4 and n(A $\cup$ B) = 8.
Venn diagrams can show the number of elements instead of the actual elements.

Use Venn diagrams to find probabilities: E.g. $P(A \cup B) = \dfrac{n(A \cup B)}{n(\xi)} = \dfrac{8}{10} = \dfrac{4}{5}$

Sampling and Data Collection

Definitions of Sampling Terms

POPULATION	The whole group you want to find out about.
SAMPLE	A smaller group taken from the population.
RANDOM SAMPLE	Every member of the population has an equal chance of being in the sample.
REPRESENTATIVE	Fairly represents the whole population.
BIASED	Doesn't fairly represent the whole population.
QUALITATIVE DATA	Data described by words (not numbers).
QUANTITATIVE DATA	Data described by numbers.
DISCRETE DATA	Data that can only take exact values.
CONTINUOUS DATA	Data that can take any value in a range.

Choosing a Simple Random Sample

1. Give each member of the population a number.

2. Make a list of random numbers.

3. Pick the members of the population with those numbers.

Random numbers can be chosen using a computer/calculator, or from a bag.

Spotting Bias

Two things to think about:

1. When, where and how the sample is taken.

2. How big the sample is.

- If any groups have been excluded, it won't be random.
- If it isn't big enough, it won't be representative.
- Bigger samples should be more reliable.

Estimating Population Size

Use a capture-recapture method:

1. Take a random sample of a population, tag them and release them.

2. Take a second random sample later on and record the fraction that are tagged.

3. Assume the fraction of tagged members in the second sample is the fraction of tagged members in the whole population.

EXAMPLE

An ecologist catches, tags and releases 10 badgers in a forest. She returns 2 weeks later and catches 15 badgers. 2 of them are tagged. Work out an estimate for the population of badgers in the forest.

1 → $\frac{10}{P} = \frac{2}{15}$, ← 2

3 so $P = \frac{10 \times 15}{2} = 75$

58

Averages and Ranges

Mean, Median, Mode and Range

MEAN	Total of values ÷ number of values
MEDIAN	Middle value (when values are in size order)
MODE	Most common value
RANGE	Difference between highest and lowest values

Arrange the data in order of size to find the median. It helps when finding the mode and range too.

EXAMPLE

Find the mean, median, mode and range for the data below:

2.4   2.8   1.7   3.4   2.6   3.6   2.4   1.9

$$\text{Mean} = \frac{2.4+2.8+1.7+3.4+2.6+3.6+2.4+1.9}{8} = \frac{20.8}{8} = 2.6$$

The 4.5th value is halfway between the 4th and 5th value.

In order: 1.7   1.9   2.4   (2.4   2.6)   2.8   3.4   3.6

Median = 4.5th value = 2.5   Mode = 2.4

Range = 3.6 − 1.7 = 1.9

Quartiles

Formulas are for a data set with n values.

LOWER QUARTILE, Q_1	The value one quarter (25%) of the way through a data set	$\frac{n+1}{4}$
MEDIAN, Q_2	The value halfway (50%) through a data set	$\frac{n+1}{2}$
UPPER QUARTILE, Q_3	The value three quarters (75%) of the way through a data set	$\frac{3(n+1)}{4}$
INTERQUARTILE RANGE, IQR	Difference between upper quartile and lower quartile (contains middle 50% of the data)	$Q_3 - Q_1$

Box Plots

The range is affected by outliers, the IQR is not.

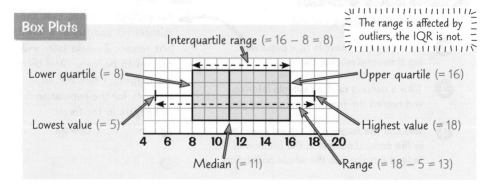

Interquartile range (= 16 − 8 = 8)

Lower quartile (= 8)

Upper quartile (= 16)

Lowest value (= 5)

Highest value (= 18)

Median (= 11)

Range (= 18 − 5 = 13)

Section 7 — Probability and Statistics

Frequency Tables

Finding Averages from Frequency Tables

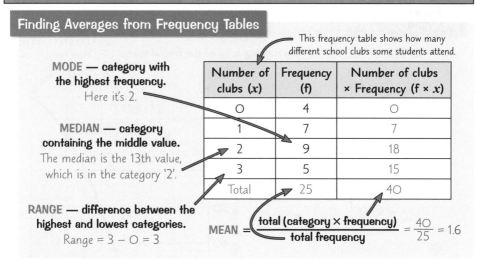

This frequency table shows how many different school clubs some students attend.

MODE — category with the highest frequency.
Here it's 2.

MEDIAN — category containing the middle value.
The median is the 13th value, which is in the category '2'.

RANGE — difference between the highest and lowest categories.
Range = 3 − 0 = 3

Number of clubs (x)	Frequency (f)	Number of clubs × Frequency (f × x)
0	4	0
1	7	7
2	9	18
3	5	15
Total	25	40

$$MEAN = \frac{total\ (category \times frequency)}{total\ frequency} = \frac{40}{25} = 1.6$$

Grouped Frequency Tables

Data is grouped into classes, with no gaps between classes for continuous data.

Inequality symbols are used to cover all possible values.

Height (h cm)	Frequency (f)	Mid-interval value (x)	f × x
0 < h ≤ 20	12	10	120
20 < h ≤ 30	28	25	700
30 < h ≤ 40	10	35	350
Total	50	—	1170

Find the mid-interval value by adding up the end values and dividing by 2.

MODAL CLASS — class with highest frequency.
Here it's 20 < h ≤ 30.

CLASS CONTAINING THE MEDIAN — contains the middle piece of data.
Median is the 25.5th value, so class containing the median is 20 < h ≤ 30.

RANGE — difference between the highest and lowest class boundaries.
Estimated range = 40 − 0 = 40 cm

MEAN — multiply the mid-interval value (x) by the frequency (f).
Divide the total of f × x by the total frequency.
Estimated mean = $\frac{1170}{50}$ = 23.4 cm

You don't know the actual values for grouped data so can only estimate the mean and range.

Cumulative Frequency

Drawing Cumulative Frequency Graphs

CUMULATIVE FREQUENCY — the running total of the frequencies.

Total number of data values.

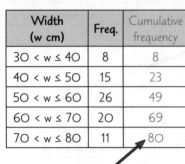

Width (w cm)	Freq.	Cumulative frequency
30 < w ≤ 40	8	8
40 < w ≤ 50	15	23
50 < w ≤ 60	26	49
60 < w ≤ 70	20	69
70 < w ≤ 80	11	80

Dotted lines are for estimating — see below.

1 Add a column and add up the frequencies working down the table.

2 Plot the points — use the highest value in each class and the cumulative frequency.

3 Join the points with a smooth curve or straight lines.

Also plot a point at the lowest value in the first class, with cumulative frequency 0.

Estimating From Cumulative Frequency Graphs

You can also estimate percentiles — e.g. the 20th percentile is 20% of the way through the data.

Go up to the value on the cumulative frequency axis, across to the curve, then down and read off the bottom axis.

- **To find the median, use the value halfway through the cumulative frequency.**
  In the example above, that's 40 — so the median is approximately 57.
- **For the lower and upper quartiles, use the values 25% and 75% of the way through.**
  Here, that's 20 and 60 — so $Q_1 \approx 49$ and $Q_3 \approx 65$. Then IQR ≈ 65 − 49 = 16.

To estimate the number of values less than or greater than a given value:

1 Draw a line up from that value on the bottom axis to the curve.

2 Draw a line across to read off the cumulative frequency.

Histograms and Scatter Graphs

Histograms and Frequency Density

Two differences between histograms and bar charts:

1 **The vertical axis of a histogram shows frequency density, not frequency.**

2 **The bars on a histogram can be different widths.**

Frequency Density = Frequency ÷ Class Width

Frequency = Frequency Density × Class Width = Area of Bar

Height (h cm)	Freq.	Frequency density
0 < h ≤ 20	16	0.8
20 < h ≤ 25	20	4
25 < h ≤ 40	30	2
40 < h ≤ 50	25	2.5

Add a frequency density column.

Use the second formula to estimate the frequency in part of a class — just work out the area of that fraction of the bar.

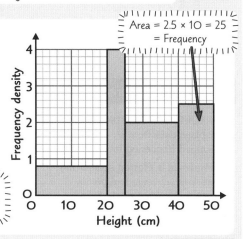

Area = 2.5 × 10 = 25 = Frequency

Scatter Graphs and Correlation

LINE OF BEST FIT — goes through or near most points. Shows correlation and can be used to make predictions.

Even if two things are correlated, it doesn't mean that one causes the other.

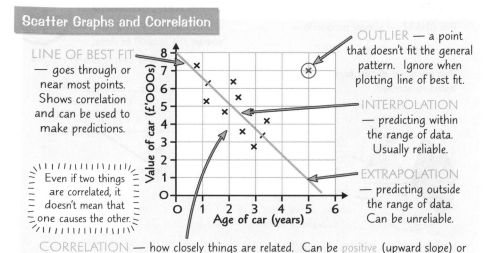

OUTLIER — a point that doesn't fit the general pattern. Ignore when plotting line of best fit.

INTERPOLATION — predicting within the range of data. Usually reliable.

EXTRAPOLATION — predicting outside the range of data. Can be unreliable.

CORRELATION — how closely things are related. Can be positive (upward slope) or negative (downward slope), strong (points close to a line) or weak (points further from line).

Other Graphs and Charts

Time Series

TIME SERIES — a line graph showing seasonality (a basic repeating pattern).

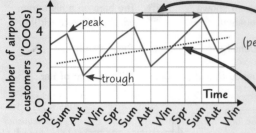

PERIOD — time taken for a pattern to repeat itself (peak-to-peak or trough-to-trough).

The dotted line shows the overall trend — e.g. here, values are generally increasing.

Frequency Polygons

FREQUENCY POLYGON — displays data from a grouped frequency table.

Frequency is plotted against the mid-interval value and points are joined with straight lines.

Weight (w kg)	Freq.
$10 < w \le 12$	9
$12 < w \le 14$	14
$14 < w \le 16$	18
$16 < w \le 18$	10

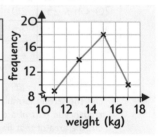

Pie Charts

PIE CHART — shows proportions.

Total of all data = 360°

This pie chart shows how 120 pupils travel to school:

The angle for the 'train' sector is $360° - 60° - 135° - 120° = 45°$.

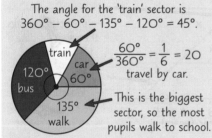

$\frac{60°}{360°} = \frac{1}{6} = 20$ travel by car.

This is the biggest sector, so the most pupils walk to school.

Each pupil is represented by $\frac{360°}{120} = 3°$.

Stem and Leaf Diagrams

STEM AND LEAF DIAGRAM — shows the spread of data.

Use them to find averages and ranges.

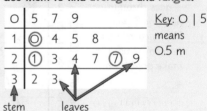

Key: 0 | 5 means 0.5 m

stem   leaves

Range = 3.3 m − 0.5 m = 2.8 m
Mode = 2.7 m
Median = 2.1 m
Q_1 = 1.0 m, Q_3 = 2.7 m
IQR = 2.7 m − 1.0 m = 1.7 m

MQHN041_MXHN041